THE PLOUGH AND THE SWASTIKA
The NSDAP and Agriculture in Germany 1928-45

J. E. Farquharson

SAGE Studies in 20th Century History
Volume 5

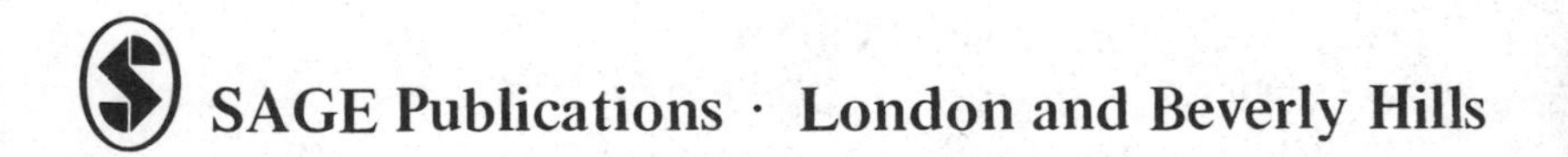

SAGE Publications · London and Beverly Hills

For information address

SAGE Publications Ltd
44 Hatton Garden
London EC 1

SAGE Publications Inc
275 South Beverly Drive
Beverly Hills, California 90212

International Standard Book Number
0 8039 9949 6 Cloth
0 8039 9997 6 Paper

Library of Congress Catalog Card Number
74-31570

First Printing
Printed and Bound by Biddles Ltd., Guildford, Surrey

CONTENTS

ABBREVIATIONS OF SOURCES

ADC	American Document Centre, Berlin.
BA	The West German national archives, Koblenz.
BM	British Museum.
DNB	Deutsches Nachrichtenbüro, a press agency.
GSA	Prussian State Archives, Berlin.
HSA(D)	State Archives North Rhine-Westphalia, Kaiserswerth, Düsseldorf.
HSA(S)	State Archives, Baden-Württemberg, Stuttgart.
IfZ	Institute for Contemporary History, Munich.
LA	State Archives Schleswig-Holstein, Schleswig.
LUD	Schloss Ludwigsburg, a branch of Baden-Württemberg archives.
ND	Darré's personal letters, circulars, etc. in the city archives, Goslar.
NSA	State Archives, Lower Saxony, Hanover.
RGB	*Reichsgesetzblatt*, the journal containing German laws, published by the Ministry of the Interior.
RMB	*Reichsministerialblatt der landwirtschaftlichen Verwaltung*, published by the Ministry of Food and Agriculture.
Unna	District Office of the Agricultural Chamber for Westphalia-Lippe, Unna, Westphalia.
VB	Völkischer Beobachter.
VJH	*Vierteljahrshefte für Zeitgeschichte*, Munich.
WB	*Westfälischer Bauer*, Catholic farm journal prior to October 1933.
WTB	Wolffs Telegraphisches Büro, a press agency.

The document referred to in the text as *Trial Brief* is in the library of the Landwirtschaftliche Hochschule, Stuttgart-Hohenheim, as are the Entscheidungen des Reichserbhofgerichts and relevant copies of *Odal* and the *NS Monatsheft*. The *Westfälischer Bauer* is in the library of the Agricultural Chamber, Münster.

J. E. Farquharson trained originally as a primary school teacher, taking a first-class honours degree in History and Education from the University of London in 1969. He took his PhD in History from the University of Kent in 1972. Since October 1971 he has been a Research Fellow, and is now a Lecturer in European History at the University of Bradford. He published (with S. Holt) *Europe from Below*, a study of recent Franco-German popular contacts, in 1975, and has contributed to several journals, including the *Journal of Contemporary History*.

FOREWORD

It must be made clear in advance that this work, which originated as a PhD thesis for the University of Kent, is not primarily an economic treatise, but rather an attempt to show the NSDAP impact upon the agrarian population, how this assisted the rise of Hitler's party to power and then what structural changes took place in the agrarian sector as a result. How the advent of National Socialism affected agriculture's place in the economy and in German society as a whole will also be dealt with. The link between foreign policy and that of autarky as an economic goal will also be discussed, not to apportion guilt or otherwise for the war, but because the two in the Third Reich are inseparable.

Some recourse to economics as such, and to statistics, cannot be avoided, since the background to the NSDAP's rise to office must be sketched in, as must the connection between economic considerations and policy-making in the Third Reich. Figures have often been relegated to footnotes in order not to disturb the narrative, but they cannot be left out altogether.

In order to show what themes will be dealt with, a series of propositions are listed here for which evidence will be offered in this work. First, it is hoped to show that the NSDAP took little interest in the land prior to 1928 except in isolated cases, and the party's programme of March 1930 and subsequent activity in rural areas was chiefly opportunism. It will be maintained here that this action was prompted by a growing radicalisation on the land, due to the economic crisis from 1928 onwards and also to the fact that the rural community had systematically been stirred up against the Weimar Republic throughout the twenties by farm unions and Right-Wing politicians prior to the rise of the NSDAP. It will be suggested that it was the peasants rather than the estate-owners who flocked to the NSDAP, so that the sociological structure of German agriculture, based on a very large number of relatively small holdings, was probably a factor in helping the party into office. The tactics of legality followed by the movement will be assessed as contributing to electoral success, in contrast to the methods of some of its rivals.

As far as actual policies initiated in the Third Reich are concerned, these were by no means always original, and it is hoped to illustrate in this work that in many respects the NSDAP was firmly in the tradition of the German Right, by goal if not by method. This is held to be particularly true of the policy of autarky, and evidence will be produced demonstrating that the Weimar Republic went some way along the road towards self-sufficiency and protective tariffs before Hitler's accession. Similarly the continuity of economic policy in the Third Reich will be stressed in so far as the question of whether priority should be given to industry or to agriculture tends to crop up again and again in Germany from the turn of the century onwards.

Furthermore, it will be held here that in purely economic terms the NSDAP aided the German farmer greatly from 1933 to 1936 to regain viability for his holding, partly through the use of the fixed-price system. But it will be suggested that the situation deteriorated for the food-producer past 1936, partly because of the introduction of the Four Year Plan, and partly because of even earlier decisions made in the history of the Third Reich which gave priority to stability of wages and prices in general. It will not therefore be maintained here that fixed prices for agricultural products were wrong in principle, but rather that after 1936 they were not high enough.

How the NSDAP accorded a special place to the peasant will be described: much propaganda in the Third Reich extolling rural virtues was pure enthusiasm in the full tradition of Romanticism: but there were solid, practical reasons for furthering the farmer as well, so that the propaganda was to a certain extent the icing on the cake only. The NSDAP genuinely believed the peasantry to be essential as food-suppliers, as the backbone of defence and above all, as the life-source of the nation: behind this credo lay the statistics of the birthrate in Germany about 1933. Ultimately, the party's belief that the rural community would always produce more children than city-dwellers was incorrect but appears to have been genuinely held. Thus in supporting and flattering the peasantry the NSDAP was romantic to a certain extent only: this should not conceal the very real politico-economic grounds for the policy.

This relative concentration upon the peasants has produced a rather unfortunate side-effect in that information relating to them tends to be more profuse than that dealing with, in particular, the East Elbian Junkers. The lack of material regarding their condition in the Third Reich is now all the more pronounced due to the presence of the Iron Curtain in Europe, behind which the former E. Elbian estates lie. It must be pointed out in advance therefore that this book is concerned rather more with the peasantry as such than with the agricultural community as a whole, although some reference will be made to the other land-owning proprietors.

As far as rural migration to the cities, often called 'land flight' by the NSDAP, is concerned, its effect upon the agrarian sector as a whole will be considered. It is a curious fact that a political movement dedicated to furthering the land should have presided over so many people leaving it. This paradox grew from 1933 onwards, as industry gradually recovered in Germany from the depression. The efforts of the NSDAP to prevent migration both by exhortation and decree, were fruitless: no one seems to have heeded the legal measures to stop the flight at all. This is a strange fact to the person accustomed to thinking of the Third Reich as a society ruled by intimidation and fear. But the truth seems to be that in so far as agriculture was concerned, the rulers of Germany had a healthy respect for public opinion: evidence will be offered here to show that this was true in respect of the law for the entailment of peasant holdings, as well as in the case of land flight. In both areas although legal measures were instituted to back up policy, they were never really applied as thoroughly as they could have been. Even in the case of the 'Battle of Production' on the land, exhortation seems to have been preferred to coercion. Prior to 1939 at least the Third Reich showed a wide divergence in several instances between the introduction of legislation and its whole-hearted application.

There is an obvious distinction between food-production and agriculture, and it should be emphasised that market-gardening, viniculture, fishing and forestry have not been included. Similarly, the efforts made to conserve food launched under the title of *Kampf dem Verderb* have not been described for reasons of space; these have also precluded any detailed account of the various bodies set up in the Third Reich to deal with the planning of land resources.

Finally every effort has been made to restrict the use of both initials (as in NSDAP) and German expressions. The criterion adopted in the latter case has been to employ English equivalents wherever possible and only to fall back upon the original where no precise equivalent exists. This applies particularly to names of official bodies (e.g. *Reichsnährstand*) or offices such as *Landesbauernführer*: in these cases it might be misleading to attempt a translation. In any case, there is the additional advantage in retaining the original word since it often obviated the need to use a whole phrase in English. Otherwise, translations have been made in all cases.

The opportunity is taken here of expressing grateful thanks to all those persons who have assisted with the composition of this work. This applies not only to the two supervisors at the University of Kent, Dr Ridley and Mr Langhorne, but also to a number of academic and professional staff in West Germany itself, as well as to those people who were kind enough to grant interviews and render assistance based upon their personal experiences of the period under review. Foremost among the latter were Drs Krohn and Reischle of the *Reichsnährstand* itself, Dr Merkel and Professor Haushofer, as well as Herr Kahlke, formerly an *Erbhofgericht* judge. Dr Günther Franz of the Institute for Agrarian History, Stuttgart-Hohenheim, allowed full use to be made of his library and records, and my thanks are due to him and his staff, who assisted in every possible manner. Among archivists who particularly helped, mention must be made of Dr Kahlenberg and Fräulein Kinder at the Bundesarchiv, Koblenz; Fräulein Tietze of the American Document Centre, Berlin; Dr Thévoc of the Prussian Archives, Berlin-Dahlemdorf; Dr Lendt of the Lower Saxon Archives, Hanover; Herr Wolke of the Landesarchiv Schleswig-Holstein; Dr Hillebrand and Fräulein von Woelke of the City Archives, Goslar; and Dr Möllmann of the Agricultural Chamber, Westphalia-Lippe who kindly allowed use to be made of the records both in Münster and in Unna. Drs Gies and Verhey both kindly allowed their dissertations to be used, and gave much appreciated advice. I must express my gratitude to Herr Joachim von Rohr, for his kindness in granting an interview. Finally, I should like to thank Dr T. W. Mason of St Anthony's College, Oxford for his advice on further source material in West Germany.

J. E. Farquharson
University of Bradford *April 1976*

1

THE INITIAL IMPACT OF THE NSDAP ON THE AGRARIAN SCENE

The actual foundation of the National Socialist movement has now been told rather too often to need any detailed recapitulation. Its essentially urban and lower-middle class nature has equally often been described at the time of its inception. Even when the movement slowly spread out into Bavaria from 1920 onwards farmers were under-represented in the membership, as later examinations have demonstrated. This does not necessarily imply that the NSDAP found no support whatever in rural areas, as clearly 'rural' and 'agrarian' are not synonymous concepts. Indeed, in autumn 1923 more than a half of all adherents lived on the land, according to a sample study based on approximately 9 percent of the then total membership. None the less, only about one-tenth of those investigated were actually engaged in farming for a living.[1] The proportion of craftsmen and tradesmen (*Handwerker*) in the party increased in accordance with the distance of their dwellings from urban centres. This fact underlines that the subsequent NSDAP advance on the land was not in all cases due to votes from farmers or peasants.

By February 1920 the party had produced a vague, generalised programme based on twenty-five points, of which number 17 dealt specifically with the question of land reform. The abolition of speculation in agrarian property was demanded, with no specific details of how this was to be achieved. Additionally, there was a provision for uncompensated expropriation, where necessary, in the interests of the community as a whole. In 1925-26 an attempt was made by the group around Goebbels and Gregor Strasser in North Germany to add some more substance to the rather bare bones of the original twenty-five points, but without success. Hitler himself declared the programme to be unalterable in February 1926, and abruptly terminated all discussion on the matter.

The reform group's agrarian proposals were nevertheless quite interesting, as they threw some light on the essentially bourgeois nature of the NSDAP.[2] It was suggested that farmers should be allowed to keep up to 1,000 morgen (250 hectare) of land as private property; any excess should be taken by the state and parcelled out into new holdings. Farm labourers were to get two morgen each as a private plot, a striking contrast to the quantity permitted to farmers. It is clear that the would-be reformers were aiming at a concept of agriculture based on the peasant holding rather than on any system of collectivised farming.

Other points included the reorganisation of marketing, based on obligatory membership of a parish sales co-operative with fixed price contracts. Even more interesting was the suggestion that new farms created through the distribution of surplus land as already described would be held on a tenurial basis from the state as ultimate landlord. The individual plots would be entailed, and ineligible as collateral against a loan. This was in line with the opening sentence of the proposals: 'Land and soil are the property of the nation.' In spite of Hitler's ban on the re-drafting of the original twenty-five points, the ideas had an unconscious predictive value. They foreshadowed in principle the actual legislation carried out after Hitler's accession, with its emphasis on combating speculation and the free play of market forces on the land, and the transformation of the peasantry into a neo-feudal estate of the realm. However flexible and pragmatic the electoral tactics of the party were on the land, its fundamental attitude towards the peasantry remained unchanged.

This was grounded in two suppositions; first, the agricultural community was conservative in outlook and therefore represented a socio-political bulwark against Communism. Statements of this nature abound in NSDAP speeches and writings. Secondly, Germany should be self-sufficient in foodstuffs as far as possible. This belief was reiterated by Hitler again and again in passages which have now become familiar, where the advantages of autarky over international trade are extolled. The need to bring living-space into a sound relationship to population is a much-repeated Hitlerian theme, to which he still adhered in the Third Reich, as his memorandum in August 1936 testifies.[3] The Führer's concepts seem to have been in sum essentially Malthusian in respect of population increase and possible food production. His attitude to the peasantry was thus formed both by socio-political and economic considerations.

Incidentally, this does not imply that Hitler had any particular interest in farming in detail; his ideas remained always general rather than concrete. Unlike Mussolini, who took pains to publicise himself driving tractors and chatting to peasants, the Führer had no real feeling for the land and its inhabitants.[4] In 1937 he was given the deeds of an estate in Lower Saxony, but subsequently never visited it.[5] He had a use for the peasantry, rather than any particular sympathy.

Initially, the whole movement seemingly paid little attention to the landed population. Activity was concentrated mainly on the industrial areas, in order to win the working-class for the national idea and defeat Marxism on its own ground. Even in rural East Friesland the Gau headquarters concentrated its campaign on the few industrial places in the area.[6] As late as March 1927 a report submitted to the Ministry of the Interior described NSDAP agitation in rural areas as 'very slight'.[7] Of course, there was some, but it was locally, rather than nationally, directed and not always very successful. One branch account from Lower Saxony reported peasant allegiance in the area to the *Stahlhelm,* the Right-Wing ex-servicemen's league, and the consequent difficulty of winning the voters for National Socialism.[8]

By late 1927, however, there were some signs of an increasing awareness at party headquarters and elsewhere of the potential voting strength which a discontented peasantry had to offer Right-Wing radicals. Hitler himself made a speech in Hamburg to agrarian representatives and more importantly told the Gauleiters in December that the party had failed to win the industrial working-class. He envisaged switching to a campaign aimed more specifically at the lower-middle strata of society in future. This social category included the peasantry by definition. Propaganda on the land now began to increase quite significantly. Interestingly, a farmer, Werner Willikens, was chosen as Reichstag candidate in South Hanover-Brunswick in preference to a railway employee, in order to bring in the rural vote.[9] In April 1928 any ambiguities in the original Point 17 dealing with expropriation were officially cleared up by Hitler. It was now stated that such action would only be taken against Jewish speculators, so that the ordinary food-producer need not fear losing his farm.

Attacks on the leaders of the existing agrarian associations were intensified; they were often stigmatised as 'Freemasons' among other things. A typical NSDAP approach is illustrated in a December 1928 meeting in Niederochtenhausen (Lower Saxony). An hour and a half long speech was devoted to attacks on Jews, Freemasons, Marxism, the press, reparations and 'interest-slavery'. Significantly, the meeting was not only well attended, but a Farmers League *(Landbund)* journal conceded that most of the audience were in its association.[10] Evidently the NSDAP was already opening up a gap there between leaders and led. This does not mean, however, that a real national drive to convert the agrarian population had yet been launched. At party headquarters there was no formal organisation specially for the land, and the overall pattern of activity was variable. As late as 1929, for example, there were complaints of inactivity at Munich. After a village meeting in Bavaria the deputy Gauleiter wrote to the propaganda department at headquarters asking where leaflets for the peasants were to be found.[11]

Despite the uneven development the results of the Reichstag elections in May 1928 appeared to confirm the thesis that the NSDAP had made little inroads on the working-class loyalty to Left-Wing parties. In the Ruhr, the scene of a heavy propaganda concentration from 1925, the party polled 1.3 percent of the total vote. For the country as a whole the results were little better. But in support of the correctness of the re-orientation towards the land were some statistics from Schleswig-Holstein, where the NSDAP share of the poll had been relatively high in some rural districts, in Norderditmarschen 18.1 percent. Schleswig-Holstein is an illustration of the party's breakthrough in microcosm, and of how the NSDAP skilfully exploited the discontent stirred up by others in order to bring all opponents of the 'System', the Weimar Republic, under its flag.

Statements about the province are necessarily generalisations, as the pattern of farming varied according to local soil and climatic conditions. This in its turn meant differing political attitudes. The dairy and livestock farmers on the west coast had an allegiance to free trade, whereas the corn growers of the east rather favoured protectionism, and hence the DNVP.[12] The latter party was beginning to lose some influence in the late twenties, as shown by its comparatively poor election results in May 1928 in the province. The other main party of the Right was the DVP, orientated towards industry rather than agriculture; moreover the death of Stresemann in 1929 robbed it of its best-known leader.

A certain vacuum began to appear on the Right in an area with a record of protest, at least since 1918 heightened by the loss of North Schleswig under the Versailles Treaty. For example, a local politician, Iversen, founded the Schleswig-Holstein Peasants and Landworkers Democracy after the war, based on 'green' rather than 'golden' democracy. By this was meant a rural movement to combat the evil of modern, materialistic capitalism. It was violently anti-semitic, and demanded the disenfranchisement of all Jews. It is interesting that the movement's business manager in the early twenties was Hinrich Lohse, afterwards NSDAP Gauleiter of the province. This comparison cannot be pushed too far; Iversen was a pious man himself, and generally supported the liberal DDP in the provincial assembly. But the anti-semitism and anti-capitalism was a storm warning, although in the period of the relatively calm mid-twenties the movement died out.

The area also had its share of other bodies whose attitude to the Republic was hostile in varying degrees. In this respect the *Landbund* was nearly always to the fore. When peasants committed bomb outrages (of which more later) its newspaper openly sympathised with the perpetrators.[13] Later evidence uncovered by the police suggested that the *Landbund's* own district chairman in Bremervoerde had been directly implicated.[14] Equally active in Schleswig-Holstein was the *Stahlhelm,* with sixteen to eighteen thousand members, some of whom also participated in the bombings. Its general outlook was exemplified by the speech of a local leader, who declared that 'healthy instincts' and 'battle capacity' were more important to youth than mere academic knowledge.[15]

Militants proliferated in the area in the late twenties, to such a degree that detailed examination of each group is impossible. There were among others the German Racial Freedom Movement, the Pan-German League, the Ehrhardt Circle and the *Werwolf.* Members of all these groupings were implicated in the bomb outrages over and above those of the two movements already mentioned. Some members of the NSDAP and the DNVP also took part. There was in many cases duplication of membership in these associations so that the militancy was not quite so widespread as might at first sight appear, but the picture for a supporter of Weimar was hardly a happy one, especially as some of the groupings were so aggressive; in 1926 the *Werwolf* were summarily disbanded by the authorities in the Pinneberg district when members were caught drilling with infantry rifles.[16]

The background was agrarian depression, as revealed in the annual reports from 1927 to 1929 of the local Agricultural Chamber. In the last quarter of 1928 there were eighty-one foreclosures on agrarian holdings in Schleswig-Holstein as a whole.[17] This was a tiny percentage of all its farms, but it provided the peasantry with a solid, genuine grievance against a republic which emotionally they had perhaps already rejected. At a time of falling prices for foodstuffs it is not easy to say exactly how hard the peasants were hit, since obviously individual cases varied enormously. One police report even suggested agitation alone and not genuine need was responsible for stirring up the peasants, many of whom allegedly ran cars. The evidence of distress is conflicting; ultimately no doubt the fact that the peasants felt themselves to be hard done by was more significant in forming their views than any objective facts about price-indices. Hence the local farm union circular of November 1928 calling on all members to find every legal way of tax-evasion until the government restored profitability to agriculture.

A series of demonstrations had already taken place in the province in the previous January, allegedly attended by 140,000 people. The spontaneous nature of the actions can be seen from the remark made by a farm union leader in Berlin, when told of the event, 'we are superfluous in Schleswig-Holstein, our peasants lead themselves'. At Längenols fuel was added to the flames when a police detachment opened fire on the demonstrators.[18] Taking their cue from the peasants the local farm unions took a step towards unity on the agrarian front. On 16 February 1928 a formal fusion at Neumünster produced a new regional federation with 35,000 members.[19] Scenting danger the government of the Reich hurriedly invited representatives of the new body to Berlin to talk things over. The Chancellor, Marx, promised a respite in legal action against farmers whose debts exceeded one-third of their property's value. He took the opportunity to warn the deputation that a large-scale relief was bound to take time.[20] This was exactly what the moderates did not possess. Johannsen, chief organiser of the January demonstrations, soon lost his influence on the peasantry by appearing to be too trustful in regard to official promises. His replacements were men of a different calibre.

A new and more radical movement now arose from the stress of the times, the Rural Peoples Movement (LVB), not entirely confined to peasants, but dominated by them. Its chief leaders were Klaus Heim and Wilhelm Hamkens: the first had been in the Pan-German League, the second in the *Stahlhelm* (he had been a wartime officer), and had a farm at Tetenbüll. Initially his model for peasant action was Gandhi, by which Hamkens meant non-co-operation with the Prussian state as far as possible, clearly a more radical concept than mere demonstrations. The symbol of the new movement was a plough and sword on a black field, an ominous choice; it was under a black flag that the peasantry had revolted in the sixteenth century.

The aims of the LVB were summed up in a circular which fell into police hands. It demanded economic security for the rural community, the recognition of its cultural importance and a political state where every trade and profession enjoyed its own representation in a national assembly. This presumably was intended as an alternative to normal parliamentary democracy where the whole people acting as one body elects the national representatives. As an idea a vocationally-based assembly was hardly original and could be found as part of the stock in trade of almost all Fascist or Right radical movements of the time. The Strasser group had suggested a roughly similar arrangement in 1925-26, as part of their proposals to which reference has been made. Presumably the object of all similar concepts was simply to limit the number of working-class representatives by taking the whole of that class as one section of the population only, to have equal representation with every other one. The numerical preponderance of the workers in society would thereby be nullified.

As a movement the LVB was classifiable as a mixture of Fascism and Poujadism. The latter aspect came out clearly in its statement that civil servants must be taught to respect the sanctity of private property; this was undoubtedly a reference to state action against peasants or others in arrears with taxes. That the LVB could also demand a recognition from the government of the cultural importance of the rural community suggests a slight inferiority complex, based perhaps on the feeling that modern developments seemed always to imply urbanisation, and consequently less attention to country dwellers as such. The LVB manifesto is therefore a useful guide to the feelings of the peasantry at a time when the NSDAP appeared on the scene as a sworn opponent of the 'System' which apparently neglected them. Numerically, of course, the rural community was in decline, which provided its members with the raw material of their grievances. After that, high taxes, falling prices for food and government assistance to export industries merely added fuel to the flames..

Even more indicative of the Poujadism of the LVB was its political naivety in practice. The decision was taken in October 1929 not to present any candidates at political elections; more direct action was sought. Before its course was fully run the LVB stirred up the peasantry to a pitch of hitherto unknown resentment, of which the NSDAP, with its greater political acumen, was the almost inevitable beneficiary. Historically the LVB had two important aspects; first, as a reflection of rural feelings and attitudes towards the Republic; secondly, that having sown a field with violence and demagoguery, it abandoned the harvest to more able hands.

Exactly how it added to unrest in the province is clear from an examination of its activities. Initially the movement intervened at auctions of farm property in order to prevent them, where they were the result of legal action against distressed peasants. Between April 1928 and March 1929 twenty-four such proceedings out of ninety-four were affected in this way.[21] The provincial government had to institute police protection at auctions, and decreed that all peasants taking part should surrender any hunting weapons at the door when they arrived.

From such actions the LVB escalated to tax strikes, in accordance with Hamkens' concept of non-collaboration with the state. Already fearful of the movement, the government now had to take more positive steps and Hamkens was arrested. It was indicative of the rural mood that hundreds of peasants attacked the courtroom where he was appearing and later stoned the police.[22] The LVB leader had incidentally already sent a letter to parish council chairmen in the Eiderstedt district, asking them if they were prepared to inform the authorities that peasants could no longer pay taxes without falling victims to Jewish capitalism. In February 1929 the tax office confiscated money due to Hamkens as part payment of his own arrears; he promptly turned up at the bureau accompanied by eighty to one hundred supporters. They forced their way in and refused to leave. The LVB leader was again taken into custody and charged with breach of the peace. He responded by way of a manifesto issued from gaol calling for 'no positive co-operation for this system and its hangers-on'.

Parallel with all this ran the LVB's campaign of violence. Bombs were placed in inland revenue buildings or those of other branches of the administration; twenty-four persons were arrested on suspicion of complicity, including both Hamkens and Klaus Heim. When the latter was released two hundred peasants staged a demonstration, which ended in fighting with the police, and a subsequent rural boycott of Neumünster, where the event took place.[23] Hamkens continued his propaganda campaign in autumn (the actual bombings appeared to have been Heim's responsibility, for which he got seven years imprisonment). These were often quite well attended, but the fire had gone out of the movement. By October 1929 Hamkens in desperation tried to work out an alliance with General Ludendorff's extremist *Tannenberg Bund,* but it came to nothing and the LVB passed virtually into oblivion. Heim was approached in prison both by the NSDAP and the communists but he declined the offer in both cases.[24] After serving his sentence he disappeared from political life. Others were ready to step into the breach. The NSDAP did not commence its activities merely on the demise of the other movement; it had begun a drive on the west coast in late 1927, where 150 meetings were held in the last quarter of the year.[25] By autumn 1928 opportunities for advance in the province were so obvious that the new Gauleiter in East Prussia was told by way of emphasis that it was as potentially fruitful as Schleswig-Holstein.[26] Indeed, the movement was flourishing, hence Hitler's speech at Heide in October 1928: this was the signal for an even more intensive drive. The party claimed to have held 5,000 meetings in that year in the province. By February 1929 the provincial administration was remarking on the scale of NSDAP agitation.

Increasing regional interest turned into a real breakthrough for the party after an incident at Wohrdener in the following month. Communists clashed with party members in a brawl and two National Socialists were killed. This was a real chance to make an impact on the west coast by building them up as martyrs in the cause of the fight against the Republic and its allegedly attendant evil of Marxism. The full range of the propaganda resources was employed at the two funerals, at which both Lohse, the provincial Gauleiter, and Hitler were present as orators, accompanied by three to four hundred uniformed members and an SA band from Hamburg. Two thousand spectators attended the first interment and double that number the second. The Führer bade a solemn farewell to his fallen comrades, victims, as he put it, of their inability to adjust themselves to the 'System'. This was by definition the fault of the latter, not of the National Socialists.

The immediate effect was a rapid conversion in the whole district of large numbers of rural inhabitants to the NSDAP. At nearby Albersdorf membership went up by over one-third in the following week. The Hitler movement was now being accepted as the collective organisation for all those who were against the Republic. By the time of the annual provincial party meeting in April 1929 Lohse could report that solid progress was being made; it is significant that nearly one-quarter of all enrolled persons had joined since 1 March that year. The party's growth was accelerating. A factor here was the close-knit rural society, which induced pressure from existing peasant members on others in the parish still outside, which in some cases amounted to the threat of a boycott if they did not join in.[27] Evidence of similar rural trends have been reported for Bavaria, where in one sample study in 1923 there were almost no cases of peasants joining as individuals. When they did enrol at any one branch on a particular day they came in groups.[28] The social homogeneity of the village community was in itself advantageous to a growing party.

Membership did not in any way imply any particular political sophistication. The police themselves said that the peasants had no conception of what National Socialism really meant; in Ditmarschen many thought that the party journal *Tageszeitung* belonged to the LVB. This must have facilitated the crude NSDAP propaganda such as attacks on 'interest-slavery', especially at a time of real economic depression. As one later NSDAP account stated frankly: 'the movement would never have conquered the people . . . if its propaganda speeches had been based upon reason only'.[29]

The other radicals of the Right began to wilt, as Hitler drummed in all the discontented under his banner. This is made crystal clear by police and local authority reports for the province for 1929. By February the DNVP was said to have neither the tight organisational structure of the NSDAP nor its available finance, and could not compete. Equally, the *Stahlhelm* began to feel the pinch; by the spring of the same year it had almost disappeared at Itzehoe, where the NSDAP had swallowed it. Three months later the Regierungspräsident told Berlin that both it and the *Tannenberg Bund* were now receding so rapidly that there was no longer any point in mentioning their activities. A movement called the *Wachverein* associated with the LVB went over to the Hitler party at Buesum. The same applied to the LVB itself, for example, at Niebull: an NSDAP branch there had fifty members in February 1930, every one of whom had formerly been in

Hamkens' movement.

The absorption was a national trend. As early as March 1929 the *Stahlhelm* was beginning to feel the chill wind of NSDAP competition in Lower Bavaria as well. The Gauleiter reported that members both of the ex-servicemen's association and of the DNVP were asking him for material so that they could recruit for the NSDAP at their own meetings. Eventually the same phenomenon also occurred in Lower Saxony.[30] The organisational superiority and greater tactical skill of the Hitler movement enabled it to annihilate its Right-Wing rivals one by one, and consolidate the once-fragmented anti-republican vote.

Nothing illustrates so clearly the political skill of the NSDAP than its treatment of the LVB, whose agitation could clearly benefit Hitler. None the less, he remained wary. He even told the editor of the party organ in Lower Saxony that the march of events in Schleswig-Holstein should be braked. The risk of indiscriminate violence in the province rebounding on to the NSDAP was doubtless uppermost in his thoughts. Lohse made the same point later, when he warned that the 'stupidities' of the LVB were a 'gigantic danger' to the NSDAP.[31] The party had had previous experience in 1922, when the assassination of Rathenau, of which it was innocent, was used by the Prussian government as grounds for banning all Right-Wing radical groupings. Having eschewed the path of illegality, Hitler had no intention of being compromised by those who had not.

Consequently in summer 1929 he forbad all party members to have anything to do with the LVB. This ban was seemingly neither popular nor always obeyed. In October there was a joint LVB/NSDAP meeting in Itzehoe. But the political flair of Lohse and Hitler was nevertheless amply justified by results. By the same month the tide was running heavily against the LVB and in favour of the NSDAP, exactly because the opinion on the land was that the latter fought with 'realistic means' as compared to the violence of the former.

Confirmation was provided in the September 1930 Reichstag elections, when the NSDAP's share of the poll in Schleswig-Holstein was seven times as great as in 1928. Interestingly enough its percentage gain was similar to the losses suffered by the DNVP and the DVP combined. It would appear that the Right was becoming consolidated rather than remaining fragmented between DNVP, DVP and the völkisch groups as hitherto. Moreover, in communities of under 2,000 population the Hitler movement polled 35 percent of the vote, against less than a quarter in urban areas, which emphasises its rural base in the province.[32] However, different areas displayed marked variations; in one case two neighbouring rural districts recorded 25 and 39 percent for the NSDAP, respectively. Leadership of the local branches was presumably a factor here, so that generalisations are suspect. But of the rapid advance in the province as a whole there can be no doubt; in July 1932 this culminated in the first ever absolute majority for the NSDAP in any constituency, when it polled 51 percent of the total votes cast.

Even this needs some qualification as there seems little doubt that the party did relatively better on the west coast than on the east, where the farms were bigger, and above all, the depression was less severe. Prices for dairy products fell relatively more than did those for grain, as informed opinion pointed out at the time. Confirmation that the peasant, rather than the estate-owner or landworker was the main NSDAP support is provided by a contemporary sociological study based on twenty selected areas. A clear correlation between farm size and voting practice emerges: owners of holdings up to twenty Ha were strongly orientated toward the NSDAP both in 1930 and two years later. Those possessing between 20 and 100 Ha were less liable to choose the party. The affiliations of the larger-scale farmers are less easily determinable, as they were bracketed in the survey with small-holders having less than two Ha each. Landworkers, not surprisingly, had a certain leaning towards the SPD.[33]

In sum, the people who voted for the Hitler movement in Schleswig-Holstein in 1930 and again two years later were more likely to be rural dwellers than urban, and more specifically to be peasants rather than agricultural labourers. This apparently confirms the picture of a lower-middle class electorate for the party, radicalised by the depression. In fact, the middle classes displayed a rather erratic pattern in voting preferences in post-1918 Germany. After swinging between DNVP and DVP in the early and middle twenties they apparently became ab-stentionists in large numbers at the 1928 polls, perhaps because some members of the DNVP supported the Dawes Plan.[34] Those who now came forward for the NSDAP were those members of the middle classes, especially in rural areas, who had been always antagonistic to the Republic but now found new occasion to desire its downfall in the economic catastrophe and fresh hope that it would be overthrown in the shape of the NSDAP. It seems ironic in retrospect that a party so urban-orientated to begin with should have achieved so sudden a breakthrough on the land.

2

THE 1930 PROGRAMME AND THE BUILD-UP AND USE OF THE AGRARIAN CADRE BY DARRÉ

At the level of the higher echelons growing interest in the land manifested itself in two articles in the NSDAP *Year Book* in 1929 and 1930 respectively, both devoted to agriculture, and both written by Werner Willikens, the party spokesman on agrarian affairs in the Reichstag. The first was couched in general terms only, the Republic being attacked because it paid too little attention to making Germany self-sufficient, a long-standing NSDAP propaganda theme. Above all, in accordance with the general slogan of 'Common good before individual gain' it was demanded that agriculture must be organised to serve the whole nation.

In the 1930 issue Willikens became more detailed and precise. First he refuted yet again the allegations that the NSDAP was a Marxist party based on expatriation. Clearly what the article described as 'lying attacks' by opponents on Point 17 were still a cause for concern. None the less Willikens emphasised that speculation in land would be ended if the party achieved power. A large settlement programme in east Germany was promised; this would be carried out at the expense of the big estates. Soil taken from them would be distributed to farmworkers aspiring to rise and to farmers' sons unable to find a holding elsewhere. All land ownership was to be confined to Germans. More particularly Willikens stressed the need for a reform of the inheritance laws. This was to ensure that no further fragmentation of holdings took place, which might have affected their economic viability. There was to be no 'land-pawning', which meant that the owner of a farm should not be allowed to dispose of it freely or use it as an object of speculation just because it was his. He had an obligation to the community as such (cf. the Strasser proposals). Despite the frequent vagueness of the rhetoric used the underlying neo-feudalism showed through quite plainly.

In March 1930 came a further step forward with an official programme for agriculture. As Willikens was still agrarian spokesman and the programme was in effect merely an expansion of his articles it seems very likely that he was mainly responsible for the text, which was divided into three sections. The first was a general review of agriculture and underlined that self-sufficiency would save imports, make for independence politically and help domestic industry by giving farmers more purchasing power, as a result of their increased output. The importance of the peasantry lay, however, not merely in its economic utility. It was the backbone of national defence and the 'life-source' of the people. This latter point referred to the fact that the birthrate was higher on the land than in the urban centres.

Section 2 dealt with the shortcomings of the current government. Taxes were too high, fertiliser and electricity too expensive, wholesalers and retailers profit margins were excessive and there was inadequate tariff protection against cheap foreign food. This package was cleverly put together; it was a long-standing complaint among farmers in the twenties that fertiliser cost too much, indeed even the semi-official Agricultural Chambers had been known to allege that a cartel appeared to exist in the industry.[1] Equally, wholesalers margins were another perennial grievance. Von Kalckreuth of the *Landbund* once said that he felt ashamed every time he tipped a waiter because the money probably represented more than the farmer had got for growing the food originally.[2] On this type of propaganda peasants were reared in the twenties. Moreover, virtually all the policy defects listed above in the programme were linked to anti-semitism, as Jewish capital was said to be behind taxes, fertiliser prices and dealers profits. There is no reason to suppose that anti-semitism was a poor card to play. In any case, the relatively heavy taxes, compared to those of 1914, was another sore point among farmers, who consistently grumbled about what they had to pay.

In the third section the party turned to its own promises for the future, reminiscent of Willikens' articles, which need not be repeated. There was, however, an apparent change of course in respect of settlement. Land for this was to be acquired by foreign policy. Willikens had foreseen taking it from existing landed estates. From the standpoint of internal agrarian administration this meant that a mixture of farm sizes was accepted, and the economic importance of large estates was recognised. In comparison to the 1930 article this programme was socially a rather more conservative document still.

In conclusion the party called for the advancement of the peasantry culturally and its social standing (cf. the LVB circular). Landworkers were to be assisted and rural migration prohibited. This combination would obviate the need for foreign labour in Germany, which would be forbidden. In fact, foreign labour was already being run down by the government anyway, which reduced the number of permits for immigrant labourers in agriculture from well over 100,000 in 1928 to 7,000 four years later.[3] NSDAP policy on this score had little practical value. As a final flourish, it was declared that only an economic recovery for the whole nation could possibly guarantee prosperity for the peasants. This was a hit at the farm unions, especially the *Landbund,* which continually demanded aid for agriculture irrespective of Germany's financial position, a standpoint which the NSDAP correctly judged to be irrational. Only a political movement aiming to destroy the system could aid agriculture in the long run. This line represented a clear recognition by the NSDAP that it had to loosen the *Landbund's* hold in order to break through on the land.

As an appeal the March programme had something for everyone, from estate owner to farm labourer. In some parts it was vague, in others surprisingly frank. But in respect of inheritance law reform and an ending of speculation it seems in retrospect to have been a statement of genuine intention.

The next step was to build up a full-scale apparatus on the land. In 1930 the necessary man was found for this, Richard Walther Darré.[4] Like several other prominent members of the völkisch Right he had been born outside Germany, in his case in Argentina. After his education in Europe (including a year at King's College, Wimbledon) he fought as an artillery officer on the Western front and afterwards qualified as an agronomist (*Diplomlandwirt*). He described himself as a strict Mendelian and specialised in animal breeding. He quickly joined a back to the land movement of Rightest tendencies, the *Artamanen,* where he met Heinrich Himmler, the movement's leader for Bavaria and himself an agronomist by professional training. Darré seems to have acquired a certain reputation in his chosen field and in 1927 was suggested by the Agricultural Chamber in East Prussia as envoy to an agricultural exhibition in Finland. The following year he worked for a time at the German embassy in Riga but was dismissed in 1929.[5] He had already involved himself in völkisch politics by joining the Stahlhelm in 1922 and later the Nordic Ring, which had links with Alfred Rosenberg's Combat League for German Culture. Possibly it was because of these affiliations that he lost two official jobs, although in 1929 he gave the reason as (unspecified) 'personal grounds'.[6] In 1928 he took up contacts with the NSDAP in Wiesbaden, but stayed outside the party officially.

His first major published work was an attempt to prove by appeal to historical criteria that the Nordics had been the true creators of European culture, in contrast to such nomads as the Jews, who were parasitical on that created by others.[7] He had been spurred on to this work by the publication the previous year of a book which had depicted the Nordics as warlike nomads themselves. Despite Darré's Lutheran ancestry he was markedly anti-Christian and blamed the 'Judo-Christian materialism' introduced by Charlemagne for the servitude imposed on the once free Germanic peasant.

His second book was written in 1929 to disseminate the idea of a new ruling-class rooted in the agrarian community.[8] Darré escalated here from breeding in the animal kingdom to similar concepts applied to human beings. Progress could allegedly best be achieved by restricting the increase of the less-valuable members of the community. Included in the book were demands for inheritance law reforms and an end to speculation in landed property, which made him obviously close to the NSDAP official line. His new peasant governing-class were to be made economically secure by the provision of entailed holdings *(Hegehöfe)*, a concept which later served as a model for legislation in the Third Reich itself. Like all Right-Wing radicals he prized the agrarian population as a force for social conservatism and believed that 'if Marxism is to be overcome in Germany the flag-carrier in this battle will be the German peasant'.

In another völkisch grouping, the Saalecker Circle, he became acquainted with Georg Kenstler, editor of an *Artamanen* magazine called *Blood and Soil.* Kenstler was yet another foreign-born German, hailing from Transylvania. He had contacts with the LVB and had been a guest at Hamkens' farm.[9] His enthusiasm for the movement led him to see in it the germ of an organisation to fight the Republic; Kenstler's idea was to build a nationwide network of cells among the radically-minded peasants, centrally controlled from Weimar: in June and

September he tried successfully to involve Darré in the project. The two men then took·up contact with Dr Ziegler, editor of an NSDAP newspaper in Thuringia. Their suggested apparatus would be financed by the party and led by Darré, who would pose as a veterinary inspector in order to allay suspicion as he went round the country.[10]

Interestingly enough the party headquarters at Munich showed a cool reaction to the whole scheme so that in March 1930 negotiations were still dragging on. The element of illegality probably was a decisive factor here; the NSDAP still sought the constitutional path to power. Darré had by now managed to get in touch with the Führer more directly by this time anyway, through the medium of another Saalecker Circle member, Professor Schultze-Naumburg.[11] Hitler had apparently heard of Darré, although he had never read any of his books, and could only offer him the post of a travelling speaker in Pomerania. Schultze-Naumburg continued his work as intermediary and in May Hess told Darré's publisher that the Führer now laid great store on acquiring Darré as an agrarian adviser. Perhaps the contact-man had discharged his duties well or perhaps Hitler had read Darré's books in the interim, since the change of attitude from indifference to warm approval had certainly been rapid. In July, the accolade was bestowed in the shape of a direct commission from the Führer appointing Darré as the leader of the party's new agrarian cadre. That older party members such as Willikens could have been passed over in favour of a newcomer seems to suggest that Hitler was highly impressed by Darré's credentials.

The new adviser was put initially under the command of Hierl of chief department II at party headquarters. His first act was to lay two memoranda before the leadership in August 1930. The primary suggestion centred around the possibility of bringing all food-producers under NSDAP control, so that a blockade of the urban areas could be organised in order to destroy the Republic; as in the case of the original project with Kenstler, Darré's ideas were still revolving around illegal methods of obtaining power. When this document was composed the party still only had twelve seats in the Reichstag, although Darré did forecast an increase to sixty in the forthcoming elections in September. In the event the NSDAP won 107, so that in effect the first memorandum was outmoded almost as soon as issued.

The second memorandum had more long-term importance.[12] In it Darré suggested how an agrarian cadre should be built up, based on the assumption that only professionals who knew agriculture could really persuade the farmers. Since the party had already advanced rapidly in rural areas without such a cadre the proposition seems doubtful in retrospect. Nevertheless, taking this as his point of departure Darré outlined a structure consisting of men who would know agriculture inside out on the one hand and be firmly under party control on the other. Every regional and local leader would have a member of the cadre to assist him in agrarian affairs, who would be simultaneously responsible to him and to the adviser's own immediate superior in the cadre, led by Darré himself from Munich.

These proposals were accepted and Gauleiters were ordered to install advisers by 1 October. There was one slight difference made to the original scheme in that 'reliable farmers' were to be taken on as honorary advisers rather than full-time professional experts, purely in order to save expense. A pyramidal structure was to be erected with advisers (LVL) in the village community, who were not required to be professional farmers, unlike those at district (LKF) and Gau (LGF) level. Appointments were to be made at all levels by the next highest adviser in the hierarchy in consultation with the political leadership on the same plane. In some cases there was an intermediary (LBF) between district and Gau and Gaue themselves were often grouped together under a regional adviser (LLF). In February 1933 there were only eleven of these for the whole country, the one for Prussia being Werner Willikens. The hierarchy was further complicated by the addition of two other posts, LAF and LOF: unlike their comrades with purely consonantal abbreviations these latter were part of the agrarian apparatus only, and had no responsibility to party political leadership at any level.

A bureaucracy had been created which by spring 1932 had become a party department (No. 5) in its own right. Growth elsewhere paralleled that at the top; in the same year Gau Saxony alone had thirty-four LKFs and 1,100 LVLs, plus forty speakers on agricultural matters, a press secretariat and a large office staff.[13] Darré had come a long way in two years. In December 1932 Hitler took the whole agrarian appratus out of the normal party system altogether and made it directly responsible to himself. This was part of an administrative shake-up engendered both by the crisis of his relations with Strasser, and by the need for greater flexibility in party organisation. The same decree gave Darré, not Goebbels, supervision of the NSDAP's agrarian journals (of which the most important was the *NS Landpost*), so that he had in effect his own complete private domain.

The build-up of the apparatus had, however, a relatively slow start. By the end of 1930 only fifteen Gaue had erected the desired structure, mostly the northern and eastern ones. Others were at various stages, including three 'dead' in terms of agrarian activity. Nevertheless they all had at least an LGF and it was decided to call a policy meeting at Weimar, after which there would be a general demonstration, to which members of agrarian associations not even in the party would be invited. Darré was well aware of the propaganda value of any such occasion, especially as some sections of the peasantry were still not convinced of the purity of NSDAP intentions regarding their holdings. Here was a chance to persuade the sceptical; moreover, the DNVP, linked with the *Landbund* by tradition, was in trouble over rifts concerning Hugenberg's leadership. Additionally, the whole agrarian community was dissatisfied with the government. Circumstances seemed to be conspiring to present the party with a unique chance to make a real impression on the waverers. Consequently Darré urged the Führer to attend in person as a propaganda exercise. On his subordinate's prompting Hitler took the chance of delivering a speech in which he promised that 'The Third Reich will be a peasant Reich or it will pass like those of the Hohenzollern and the Hohenstaufen'. These words were frequently to be cited by the agrarian cadre and presumably made some impression on the non-NSDAP visitors at the rally.

From propaganda the apparatus moved swiftly to operations against the largest farm union, the *Landbund*, which was undergoing a leadership crisis at the time. Schiele, its president, became Minister of Agriculture under Brüning in spring 1930 and broke with Hugenberg and the DNVP. Schiele entered another party (CNBLVP) and laid down the union presidency in October 1930. This caused some uncertainty in the *Landbund* ranks, as members were not clear whether to follow Schiele's party or Hugenberg's.[14] Dissension among the *Bund* leaders themselves accentuated the confusion; at national level the union declared itself behind Schiele's policies as Minister of Agriculture in April 1930. Yet in the following July a Hanover *Landbund* leader was attacking Cabinet measures as inadequate. The same month the VB was gleefully reporting how poorly-attended *Landbund* meetings were in Saxony due to the 'catastrophic policy' of the national leaders.

On 22 July the central committee of the union decided that all parliamentary action on its behalf would be carried out in future via the

CNBLVP.[15] This caused a terrible row in north and east Germany at the union's regional leadership level; Pomerania flatly refused to support any one party, according to the NSDAP press. A couple of days later the same source reported that one district branch in Reichenau (Silesia) had simply withdrawn itself from the *Bund* as a result of the central committee's decision. By late 1930 the whole organisation was in disarray, and in no position to withstand systematic infiltration by any group so organised and determined as the NSDAP. So when in November 1930 Darré ordered his new apparatus to undertake the task of penetrating into the *Bund* in order to undermine its leadership this tactic clearly had a good chance of success; he continued to advocate a similar policy of 'nibbling away' at the *Bund* throughout 1931. In particular in September of that year he urged the cadre to penetrate the *Bund* and 'conquer one position of power after another'.

The object of this exercise was not to destroy it as such but rather to take it over as the foundation-stone for the eventual NSDAP agrarian organisation in the Third Reich: this is quite clear from Darré's own directives. By June 1931 Hitler was publicly appealing to party members to join this 'great organisation, which will be well able to collaborate in the Third Reich': the *Bund* was to come under NSDAP control.

Success came swiftly in terms of winning influence within the *Bund* ranks, when the entire management committee of one Kreis branch, formerly in the CNBLVP with Schiele, resigned to make way for NSDAP members. In February 1931 some NSDAP members were elected to office for the *Landbund* committee in Silesia.[16] Two months later in Baden the LGF of Darré's apparatus was telling subordinates that the time had come to roll up the *Bund* from underneath in order to bring it firmly under NSDAP control. By September its leaders in Franconia were approaching the NSDAP with a view to collaboration. Even more important, partly due to Werner Willikens, himself a member of the *Landbund,* two important *Bund* leaders had been won for the NSDAP cause. These were Hartwig von Rheden-Rheden in Lower Saxony (later to be an important agrarian official in the Third Reich) and the vice-chairman of the Brunswick branch of the *Landbund.*[17]

Infiltration was not the only weapon used; strong attacks were also made upon RLB leaders in personal terms. When one of the latter was incautious enough to praise the government, he was promptly described in the NSDAP press as 'a friend of fulfilment', that is, of the terms of the Versailles Treaty and of Marxism. Darré himself justified the frequent practice of referring to *Landbund* leaders opposed to the NSDAP as Communists on the grounds that their non-co-operation with the party would contribute to Communist success. This convenient logic allowed the NSDAP to attack anyone attempting to obstruct its onward march.

The NSDAP also tried to force a split between the paid union officials and the elected leadership; the former were invited to come over into the NSDAP which, combined with infiltration into the rank and file membership, would produce a situation where the existing leaders were totally isolated from everyone else within the union. Officials were to be decoyed with promises of a secure future for them in the NSDAP ranks.

The undermining at one level was paralleled by defection at the top. In the late summer of 1931 one of the two national directors, von Sybel, took up contact with the NSDAP, which he did not publicly announce.[18] Darré consolidated by demanding parity of leadership at national level, on the grounds that the NSDAP now had the majority of the membership in the union. He told the president that if the leading positions were not shared, the peasants would become infuriated and destroy the whole union. In a directive Darré emphasised the advantage of gaining official representation at national level; this would stop the DNVP from using the union against the NSDAP, and secondly, pull the ground from under the feet of those who maintained that the party was not 'agriculture conscious'. Evidently NSDAP opportunism on the land was still fairly obvious.

On 8 December the *Landbund* national leaders held a conference in Berlin to discuss the current situation. NSDAP pressure received its reward when the management committee elected to yield and chose Werner Willikens as a member. The party was now well and truly on the road to total conquest, and Darré sought to hammer home the advantage gained by telling the LGFs to intensify the drive for management positions at lower levels. By February 1932 the entire leadership in Kreis Einbeck (Lower Saxony) was in NSDAP hands. In East Friesland the DNVP was complaining that local Kreis committees were coming more and more under NSDAP domination.[19]

The party could now use the union as a mouthpiece of its own propaganda, as in Saxony in September 1931. The then regional president (who was in the CNBLVP) wanted the branch publicly to support government policy. The LGF in Saxony then moved into action, first by rallying all branch members who were also in the NSDAP to act against this suggestion. He then wrote an article violently attacking the president on personal grounds as being weak and dishonourable. NSDAP lobbying was successful, and endorsement of Schiele's policies prevented. In August 1931 von Sybel's interview to *Der Angriff* stating that the *Bund* no longer had any confidence in Schiele or Brüning must have seemed all the more convincing, as no one yet knew that he was in the NSDAP. By October of the same year the *Bund* had joined the Harzburg Front against the Republic. At the second ballot for the Presidential elections in April 1932 it officially supported Hitler's candidature.

Apart from its actions against the *Landbund* the NSDAP had to reckon with the need to deal with the only truly agrarian political movement in Germany, the CNBLVP (Christlich-Nationale Bauern- und Landvolk Partei). The party had strongly attacked the NSDAP because it advocated one thing on the land and often quite another in the cities; for example, in urban areas it had criticised price increases in food.[20] Consequently, a powerful propaganda campaign was opened against the CNBLVP by the agrarian cadre and the NSDAP press in late 1931. The party was pilloried as 'Jewish-liberal' and as the slave of 'red corrupters'. In its action the NSDAP was supported, strangely enough, by the *Landbund* which doubtless found the CNBLVP with its nineteen members in the Reichstag a rival to itself as well. The CNBLVP press, especially in Thuringia, warned the Farmers League of the danger it was running in co-operating with the NSDAP, but in vain. Collaboration continued and with marked success. By November 1932 not a single CNBLVP member remained in the Reichstag. The whole episode reflects NSDAP tactical opportunism and propaganda capacity to a very real extent.

The second pillar to be attacked was the semi-official Agricultural Chambers. In August 1931 Darré appointed a special adviser on their organisation: two weeks later he was telling his subordinates that Chamber elections were now to be made a matter of party politics so that this 'instrument' could also be delivered into NSDAP hands. Lists of NSDAP candidates were to be drawn up, despite opposition from some Gauleiters who still wanted to work with the *Landbund*. For the basis of his campaign he suggested pillorying Schiele to make the farmers and landworkers dissatisfied with the Republic. He drew attention to a recent article in the NSDAP press alleging corruption in grain dealings. Financial irregularities and bad management of the Chambers themselves were used repeatedly in NSDAP propaganda activity aimed at winning control for the party in this organisational area.

To take two regional examples, the Schleswig-Holstein Chamber was in a leadership crisis, as the rank and file membership had lost confidence in their representatives. An exhibition organised by the latter in Hamburg had proved financially disastrous, losing RM 400,000: the NSDAP belaboured the Chamber President mercilessly in their journal for inefficiency. Political differences had added to the gap between leaders and led in the province, as the President was in the *Landbund* and many of the latter in a smaller association, the *Deutsche Bauernschaft.*[21] NSDAP progress was rapid in the area, and by October 1931 the party had seventy-three members in the Chamber, thirteen of whom were LVLs from the agrarian cadre.[22] In East Prussia the party followed similar tactics of personal abuse of existing leaders; the local President, Brands, was alleged to be guilty of mismanagement, as well as being a Freemason. Before the end of 1932 the NSDAP controlled fifty places out of seventy-six in management in this Chamber. Commenting on this a local newspaper stated that whereas NSDAP farmers voted en bloc in the elections, only 30-40 percent of their opponents had bothered to do so.[23] Apathy by the opposition was perhaps as great a factor in NSDAP success as their own activity at least in this instance.

The party also raised the question of contributions paid by members to the Chamber. In Westphalia the party representatives claimed success in actually lowering them from 65 Pfennigs per RM 1000 of property value to 62 Pfennigs. In addition they managed to get the (for them) unwanted President replaced and by January 1932 their own group leader, Meinberg, was on the management committee. In East Prussia also, the subscription was cut after the NSDAP won control although the decision to do so had been made previously to this; it is unlikely that many local farmers realised this.

Throughout 1931-32 the NSDAP advances continued apace in many areas of the country. By the end of 1931 the party had won one-third of all seats contested that year for the country as a whole.[24] In January and February 1932 the VB was full of reports of new successes in the Chambers, often at the expense of the *Landbund.* As was the case with that organisation, the Chambers, once captured, proved useful organs of propaganda: in East Prussia the Chamber accepted an NSDAP resolution in December 1931 recognising the helplessness of the Reichs President Hindenburg in face of the parliamentary system, and requesting that the way should be made free for men welded by 'struggle and faith' to save not merely agriculture but the entire country.

For a party possessing in March 1930 neither an agrarian programme nor apparatus, the NSDAP position by mid-1932 was astounding. A national cadre had invaded successfully the two main pillars of agrarian organisation. Even when allowance is made for apathy on the one hand and determination on the other, it seems unlikely that such an advance could have been recorded if the farmers had had confidence in the government or in existing professional bodies. This dissension obviously enormously aided Darré but it would not have sufficed on its own without the drive and the determination which the agrarian cadre had displayed at all levels. At its inception Darré had laid down that all existing agricultural associations were to be regarded as 'conquerable fortresses'. This was later expanded to the statement that 'there must be no farm or holding, no co-operative or rural industry, no local farmers union . . . where our LVLs have not so worked that we cannot immediately paralyse the structure'.

By early 1933 Darré's wish was fulfilled. In Saxony, to take one illustration, so formidable a machine had been constructed that the local leader could boast that as a result of his connections with the regional government he knew of absolutely everything that happened in agriculture in the area, and had a watchful eye on all actual or potential opposition.[25] It has to be pointed out that the policy of infiltration into the *Landbund* had not originated with Darré; in Lower Saxony it had been adopted as early as 1928.[26] Darré's arrival was therefore to be seen as the start of an escalation of existing tactics through the medium of a nationwide agrarian network.

3

THE GENERAL POLICIES OF THE REPUBLIC AND THE AGRARIAN REACTION 1928-1933

National Socialist Propaganda

The world economic crisis hit some sectors of German agriculture more than others. For example, rye declined more in value than did wheat. In fact, the latter commodity actually yielded a greater income in 1932-33 than it had four years previously. On the whole, dairy products and livestock fared even worse than grain products. Statements about the effect of the crisis on agriculture are consequently always generalisations; how the individual came out depended partly on what he was growing and partly on his own skill. Broadly speaking, the peasants did relatively badly as most dairy-farming and livestock rearing was in their hands, but even here price changes were variable: for example, milk remained relatively more profitable than cattle.[1] But against this the large-scale grain-producer sold a higher percentage of his product than did the peasant, whose family consumed some of his produce, and was less affected by a price-fall in consequence. In sum, the effects of the depression were extremely variable.

Naturally as prices fell indebtedness tended to increase.[2] Compulsory auctions and foreclosures became increasingly frequent.[3] Sometimes failure to meet obligations was due to personal incapacity rather than to sheer bad luck or the vagaries of the economic crisis. Of over 400 cases of farm bankruptcy examined in Bavaria at the time well over one-quarter stemmed from bad management, including eighty-seven cases of persistent drunkenness by the peasant concerned.[4] But it was very doubtful if angry farmers saw things in that light. All they knew was that under the Republic their incomes were steadily falling; moreover, they were in many cases already against the 'System'. In 1927 a prominent Bavarian politician told the government that local peasants were putting the blame for their post-war disappointments on to 'the parliamentary democratic system'.[5] With the economic crisis after 1928 the fires on the land were increasingly stoked up.

The NSDAP was the chief beneficiary of this process, although sometimes others performed the agitation, especially the *Landbund.* There were really two main factors to be considered apart from the economic crisis, the feeling of betrayal on the land and the divisions in the ranks of the agrarian community. No account of the NSDAP's rise would be complete without some reference to these issues.

The matter of betrayal went deep. Industrialism began relatively late in Germany so that even in 1900 academics and politicians alike were still undecided as to whether Germany's future should be as an industrial or as an agrarian state, a question that the British had decided two generations before. After the first war the obligation to pay reparations imposed on Germany the need to export at all costs, in order to acquire the necessary financial surplus. So when freedom to fix her own tariffs returned on 1 January 1925, it would have been difficult to have re-introduced high customs duties. They would have had the double effect of making food dearer, and of risking retaliation from other countries against German manufacturers. Consequently the coalition government struck a bargain and equalised industrial and agrarian duties. The influence of the Chancellor, Stresemann, was doubtless considerable because his party, the DVP, was associated with industrial interests. So grain-producers did not get the protection which they had sought, and which the DNVP and the NSDAP had advocated.[6] Another factor in influencing the government was probably the 'Fats Gap'; Germany was far less self-sufficient in fats than in grain. One reason for relatively low tariffs was to assist those peasants who produced domestic fats but imported feeding-stuffs in order to do so, and so wanted cheap foreign fodder. In this respect the new Act favoured the small farmers of the south and west rather than the eastern rye cultivators.

The debate over the Act highlighted the scissions within the agrarian community itself, where grain-producers usually demanded a far higher degree of protection than did most peasants. So whatever course the Republic followed someone would be discontented. Almost as soon as the Act was passed a wave of protest broke over the government's head. Von Kalckreuth of the *Landbund* spoke of adopting radical measures if grain-producers did not receive adequate state assistance; a local farmers union in Schleswig-Holstein demanded heavier tariffs.[7] The government produced a compromise, namely a state-financed grain-buying corporation to help maintain internal prices. Agriculture had won the first round and shown its teeth in doing so.

However, it received a setback when in 1926-27 a series of trade treaties were negotiated on favourable terms to countries such as Poland enabling them to send foodstuffs to Germany at reduced tariffs. This was to ensure overseas markets for German industry on a *quid pro quo* basis. The DNVP and the *Landbund* both carried out a shrill campaign against what they regarded as the betrayal of German agriculture in favour of the export industry. In February 1932, for example, von Kalckreuth told peasants that grain and timber were being purchased from Russia, so that exporters could sell their goods there.[8]

The split over tariffs revealed itself starkly when the Green Front, a loose alliance between agrarian associations, was formed in 1929. The object was to bring pressure on the government in the interests of agriculture. The members were the *Landbund*, the Christian Farmers Union (CBV), and its Bavarian sister association the BBB, representing the catholic peasants in the south and west and the *Deutsche Bauernschaft* a smaller dairy-farmers association. Although common demands were presented in March 1929, the illusion of unity did not last, simply because the socio-economic basis of their individual memberships was disparate. This came out in 1930 when a state maize monopoly was introduced to control the imports of feeding-stuffs. The *Landbund* welcomed this move on behalf of the eastern grain suppliers, as it offered them the chance to sell their surplus rye in lieu of foreign fodder. But as the former was dearer, this was clearly not in the interests of peasants relying on cheap imported fodder. So the *Deutsche Bauernschaft* just walked out of the Green Front. One man's protective measure was another man's higher production costs.

The Green Front was now less able to present itself as a general farm alliance especially as the *Deutsche Bauernschaft* produced its own policy statement in March 1932, *Agrarian Policy in Statistics,* which promptly drew from the CBV a pamphlet *Agrarian Policy in correct*

Statistics.[9] The waspishness of the reply is a fair indication of the ill-feeling which existed between the two associations. In essence the *Bauernschaft* wanted low tariffs, or none, on fodder; the CBV maintained that this would only lead to over-production of livestock and therefore lower prices still for the producer.

Inter-union bickering involved personalities as well as policy. A *Landbund* newspaper alleged that Dr Hermes of the CBV was receiving DM 40,000 yearly from IG Farbenindustrie, the chemical firm; as fertiliser prices were high the implication was that he was profiting from agrarian distress. A CBV-orientated journal roundly described the entire statement as a lie.[10]

Apart from quarrelling with one another, the agricultural associations also came into conflict with industry over tariffs. The government was partly to blame; in January 1930 the then Minister of Agriculture, Dr Dietrich of the DDP, declared that reparations were especially bad for agriculture, as industry had transferred the debts to farmers who could not recoup themselves by higher prices. This kind of statement reinforced suspicions about betrayal. The agrarian sector developed a positive phobia on the whole subject; a demand for free trade was described by a farmers journal as 'barefaced cheek'.[11] Readers were warned that opposition from industry to the land would grow.

In 1932 the chief German industrial confederation published its own programme for agriculture, based on better marketing and lower production costs.[12] Presumably this was to stifle cries for more protection. Higher tariffs from a manufacturer's angle simply meant dearer food and probably increased wage demands. The peasants were correct in assuming that industrial interests were often trying to deny them protection, but paid no heed to the almost certain consequences of high tariffs. At a time of mass unemployment any demands which meant higher food prices were surely socially undesirable.[13] Emotion simply seemed to have taken over. In January 1933 the *Landbund* accused the government of 'pillaging' agriculture to suit exporters. The charge was followed by an equally vehement denial from two associations for trade and industry. None the less a few days later the CBV in Westphalia passed a resolution deploring the 'one-sided furtherance of export policies'. Other interest-groups joined in the fray. When the *Landbund* demanded a moratorium on the rapidly growing agrarian debts the National League for Savings Protection immediately drew governmental attention to the thousands of letters from its members complaining that they were already unable to obtain payment on loans to farmers. The League accused the *Landbund* of one-sided demands.[14]

What could the Republic have done to fight increasing radicalism on the land? A number of policies were tried in succession, not always with success. In 1928 Schiele, the then Minister of Agriculture, introduced an emergency programme chiefly to aid the livestock and dairy produce sector, and so close the 'Fats Gap'. This plan, based on an initial subsidy of RM 60 million (later raised to RM 100 million), was a logical consequence of the 1928 Tariffs Act. Included in the programme was a subvention to pig-breeders – as their fats were used by the poor as substitutes for butter and margarine – to save imports. Schiele's successor, Dr Dietrich, continued to follow the same general line.

Then the world economic crisis began to push the government willy-nilly along the road to protectionism, as prices fell. To compound its difficulties a combination of changed eating habits and bumper harvest produced a rye surplus. The initial government reaction was to buy up 200,000 tons and so keep up prices. The measure proved inadequate, and the Green Front began to shout for fixed, minimum grain prices and import controls. Opinion was mixed on the latter concept; the SPD accepted it, but both the DDP and the DVP flinched away from the higher fodder prices which the scheme entailed. Eventually, as Dietrich said, the government yielded to the world economic situation and brought in the maize monopoly in March 1930. In a sense this was a victory for the pressure politics of the Green Front, although it led to the withdrawal of the *Deutsche Bauernschaft.* Technically the scheme was successful because in 1930-31 the total of domestic rye used as fodder was about three times as great as six years previously.[15]

Brüning's arrival as Chancellor in March 1930 saw no change at first on the agricultural front. Schiele took over again as Minister of Agriculture, a highly popular change for the peasantry. As a leading member of the Green Front and of the *Landbund* his accession apparently heralded a better deal. Hence the warm welcome extended both by the CBV and by the *Landbund* itself. In July he decreed the compulsory mixing of wheat and rye in flour for human consumption in order to take up even more of the rye surplus. All in all, the grain policy of successive governments was successful, in that German prices were insulated against the catastrophic price-falls on the world market, as a cabinet minister claimed in 1931.[16]

Given this fact, why was there so much rural discontent? The answer lay in the situation in the other sectors of the agrarian front, especially livestock and dairy farming. By 1931 cabinet policy had become dangerously unbalanced. Tariff protection afforded to grain far exceeded that for fats or live animals or meat.[17] This anomaly resulted from the nature of German trade treaties with countries such as the Netherlands or Denmark who sold livestock and dairy produce to Germany and bought her industrial goods. As these treaties were still in force political options were limited. A further factor was Brüning's attitude to customs duties: his main object seems to have been to avoid employing them as far as possible; if anyone used them first against Germany he would have a weapon with which to fight reparations demands.[18]

Hence his rejection of Schiele's proposal for butter tariffs.[19] This caused a lively reaction from peasants, which was doubly unfortunate, as in the south and west they were often his own supporters politically. Protests also arrived from areas where the peasantry were already in full cry against the Republic, for example, Schleswig-Holstein.[20] Eventually the cabinet gave way and introduced a quota system for butter imports based on 60 percent of the volume of purchases in 1931. It was unfortunate that it had to be pushed into this by public opinion. Almost certainly it reinforced the peasants' belief that they would obtain nothing without pressure. Sometimes the Republic almost seemed to invite this; in August 1930 Schiele had declaimed that there must be no future German government not influenced by the peasantry. He could scarcely complain if they took him at his word.

By the following year storm signals were flying everywhere on the land, as one agrarian association leader after another savagely attacked the Republic. In Schleswig-Holstein a recurrence of tax strikes in November heralded further trouble from that strife-torn province; the peasants were now even boycotting those who did pay. Apparently the aim was to spread this out to north and east Germany in general.[21] Both the DNVP and the *Landbund* were behind this. There was by now genuine financial distress in Schleswig-Holstein. In 1931-32 cattle deliveries from the province to Hamburg rose in volume over the previous year but yielded 20 percent less cash to their producers.[22] From Württemberg came accounts of financial collapse among peasants, due to high interest rates and lack of liquid resources. Agitation was now said to be rife 'even amongst otherwise calm people'.[23] Radicalisation was now spreading like a bushfire. Of course, Brüning was well aware of all this but existing trade treaties and a deteriorating world situation tied his hands. Delays in helping agriculture simply gave the agrarian leaders a further chance to undermine the Republican citadel. By January 1932 von Kalckreuth openly declared that no salvation could be expected from Brüning's cabinet. As the *Landbund* had welcomed Schiele less than two years previously this was a clear sign how fast the relationship between governors and governed was deteriorating. This was not so much the end of a honeymoon as a demand for separation.

Agrarian disillusionment had four main grounds. First, Schiele failed to live up to expectations. Then there was dissension in the cabinet. Thirdly, taxes remained relatively high as incomes fell: this made repayments difficult on loans, and the outcome was often foreclosure. Added to all this was the constant refrain of betrayal. But until June 1932 when reparations were lifted Germany had to sell abroad, so that agrarian demands were rather selfish in this context, hence the bitter tone of Brüning's last speech in the Reichstag. He pointed to the international crisis and unemployment in the cities. Agrarian leaders should have referred to this, he felt, and told their folowers that only a general economic recovery could aid them. Instead they had ranted about the unemployed as 'the workshy'.

In addition the whole scheme known as *Osthilfe* was a potent source of discontent on the land from 1929 onwards. This has been described so often that there is little point in doing so again. Nevertheless, it has at least to be referred to, as conducive of yet more ill-will, especially between the Junkers and the coalition Left/Centre régime in Prussia. Additionally the widespread allegations of corruption regarding the actual allocation of funds inevitably brought the Republic into disrepute. This was all the more important in that the chief beneficiaries, the eastern estate-owners, were themselves often the sworn enemies of the 'System'. Hence the agitation engendered by Osthilfe among the peasantry; it was bad enough to see public money being squandered, doubly hard to bear when so many recipients wished to destroy the government which bestowed the aid. It was consequently no surprise that disparate groups such as mountain farmers in Bavaria or the regional administration in Württemberg should express so much indignation over the whole project.[24] The *Deutsche Bauernschaft* joined the chorus in January 1933 with a resolution blaming estate-owners for the crisis on the land. Aid for them had meant less for other sectors.[25] Osthilfe was thus a major contributor to internecine warfare in the agrarian community.

As to its role in unseating Brüning, opinions differ. Hindenburg had other reasons for dissatisfaction with the Chancellor, which might have caused his downfall anyway. However, when Schlange-Schöningen replaced Treviranus as Osthilfe Commissioner he began to link the scheme with settlement projects; non-viable eastern estates he told the press 'will be used for settlement as soon as possible'. Otto Braun, Minister President in Prussia, felt this statement to be 'incautious'.[26] Junker circles around Hindenberg, notably his neighbouring landowner Oldenburg-Januschau, began to talk about 'agrarian Bolshevism'. By early 1932 Schlange-Schöningen had prepared a draft Bill with the contents of which Hindenburg was acquainted by Freiherr von Gayl, a member of his wartime staff and a landowner. This may well have played some part in removing Brüning; Hindenberg was himself a landowner at Neudeck and had been known to say that he had not won at Tannenberg in order to destroy the existing social order in the east. Incidentally, when von Papen's 'Cabinet of the Barons' removed the Prussian government from office in July 1932 the Minister of the Interior charged with the exercise was none other than Freiherr von Gayl.

The arrival of von Papen at the helm coincided temporally with the lifting of reparations at Lausanne in July 1932, which gave more scope for internal manoeuvre although Germany still had to purchase food and raw materials abroad. Nevertheless, von Papen was able to offer substantial aid to agriculture as a whole, which he conceded had been neglected in the past – a dangerous admission. Interest rates were cut by 2 percent for farmers and another RM 100 million were sanctioned for grain support. Von Papen made a *quid pro quo* to industry against these agricultural concessions.[27]

Despite Germany's relatively greater freedom in economic policy a cloud still remained on the horizon, the existing trade treaties, especially with the Netherlands. Even the quota system for butter imports had failed to clear this up completely. The Netherlands were Germany's second largest foreign customer so that anything inviting retaliation was excluded. A decision was postponed by the Papen Cabinet until the treaty expired at the end of 1932, by which time Schleicher had taken over the reins. Following in von Papen's footsteps Schleicher had already made promises to agriculture in a radio speech before Christmas. The peasants, however, noted that part of his policy was based on winning trade union support, which made them fear that fresh promises would terminate as ever in renewed 'betrayal'. He was warned by one farm journal that he would fail if he neglected the land.[28]

Schleicher, however, had to secure agreement in his cabinet in order to implement his promises, which he found impossible to achieve. The split was again over the perennial question of how to weigh off the interests of the land against those of business and trade. Freiherr von Braun (DNVP), the Minister of Agriculture, was sharply opposed by Dr Warmbold, who represented industry as Minister of Economics, and was personally associated with IG Farbenindustrie. Butter imports were linked naturally with low domestic prices; as the treaty with the Netherlands was running out, von Braun suggested that now was the time to increase the duty. Warmbold adopted delaying tactics, presumably due to his fear of Dutch counter-moves. A bitter row broke out when the Minister of Agriculture accused him of being prepared to sacrifice agriculture to the export industries. Unless a decision was reached quickly von Braun threatened to wash his hands of the whole affair.[29]

The Chancellor was toying with the idea of a compulsory mixing of butter and margarine in order to remove the surplus of the former and so keep up the price.[30] When news of these proposals leaked out it started speculation on the country's butter markets which led to a further severe drop in prices in only three weeks. Von Braun himself estimated the reductions to be costing German farmers the equivalent of RM 268 million annually. These facts, coupled with cabinet delays, loosed a storm of savage criticism over the government's head from practically all quarters of the agricultural front. Seldom had enemies of the Republic been afforded such a splendid opportunity for displaying its weaknesses.

The NSDAP press immediately exploited the cabinet divisions, proclaiming that 'the ministers quarrel and the peasant loses house and farm'. On 8 January 1933 von Kalckreuth joined in, describing compulsory mixing as unsatisfactory.[31] A local farm union in East Prussia demanded the redemption of solemn promises, and a fortnight later resolved that the fight against the government was 'a matter of honour'.[32] The chairman of this association incidentally was Oldenburg-Januschau. A *Landbund* newspaper called for a 'general offensive' against Schleicher and Warmbold. On the previous day *Der Angriff* had published eighteen petitions of protest from agrarian organisations; on 12 January, the Pomeranian branch of the *Landbund* demanded 'protection against a failing cabinet' in a telegram to Hindenburg.[33]

It is conceivable that Schleicher might have survived even this extraordinary degree of pressure had he not committed the supreme folly of quarrelling openly with the *Landbund,* far and away the largest and most influential farmers union. The association had requested an audience with Hindenburg for 11 January to explain the plight of agriculture. It was a sign of the times that the official *Bund* representatives included both Willikens and von Sybel of the NSDAP. Hindenberg was so impressed by their case that he arranged a further meeting for the evening, with both Schleicher and von Braun present. Here von Kalckreuth took the lead and painted a picture of acute distress on the land. He suggested that the government should make up its mind what it wished to protect, industry or agriculture. In view of cabinet discord this was a fair point. Schleicher replied and emphasised how difficult debt-relief for agriculture was, as farm creditors were undergoing a financial crisis too.[34] His Minister of Agriculture then wound up an inconclusive meeting by promising a more protectionist policy for the future.

But now an accident led to a total break in relations. Before seeing the Chancellor the delegates had passed a resolution attacking the government, which was handed to the press but not mentioned during the subsequent talks. Furious, Schleicher issued a statement declaring that he would not deal with the *Landbund* again. This opened the gates to a further flood of protests from its branches all over the country, couched frequently in highly emotive language. At a demonstration in Schleswig-Holstein there was talk of 'taking up the gauntlet' as though Schleicher's communiqué was almost an unofficial declaration of war. The NSDAP press printed a whole row of local and regional resolutions attacking the government in similar terms. Darré availed himself of the chance to write an open contemptuous letter to Schleicher calling for his resignation.[35]

There were certainly other reasons for Schleicher's political demise later in the month than this pressure such as the publicity given to Osthilfe and the promise to appoint an investigating commission on behalf of the Reichstag's budget committee. True, when the *Landbund* met cabinet representatives Osthilfe had not been mentioned. Nevertheless to the circles around the President a genuine examination would probably have been embarrassing. The parliamentary committee appointed did eventually issue a report based on twenty-six cases of alleged corruption. It was stated that in no single one had evidence of this been discovered.[36] It is hard to determine whether the commission was engaged in a general concealment of the facts or not. Certainly funds had been dispensed in a curious way with the lion's share going to the estate-owners. It was stated during the Reichstag's budget committee debate that the biggest 722 properties receiving aid had acquired nearly half of the funds awarded. As there were over 12,000 recipients in all it was clear that the small man was at the end of the queue. Under these circumstances the fact that peasants had received more assistance per Ha than landowners was of little consequence. Whether justice had been done or not it manifestly had not been seen to have been done.

Whether all this brought Schleicher down or not it probably contributed to his defeat and certainly to malaise on the land in general. In retrospect democracy in Germany seems to have been hit by four factors above all. First, the decline in international trade was such that its ultimate survival was problematic once the depression really began. But feeling on the land was hostile even prior to 1928; the warning had already been sounded that the first recession would play into the hands of the Right-Wing radicals and enable them to stir up trouble.[37] In the wake of a lost war and everything that that implied, there were too many people against Weimar from the beginning. Thirdly, the problem of divided counsels haunted successive coalition governments. The logic of the post-war position for the country precluded any easy answers. Germany had to export in order to earn enough to pay reparations, apart from having to import food and raw materials. Heavy duties simply could not be applied; moreover, even if they had been not all sections of the rural community would have been equally pleased, as witness the row in 1930 over the maize monopoly. Dissension was unhappily to be found as often in the branches or headquarters of different agrarian associations as in the various cabinets concerned with central policy. The general picture of discord and confusion must surely have assisted the rise of any party preaching unity in such a dynamic fashion as the NSDAP.

This the fourth point to consider. It was the energy and determination of the NSDAP that ensured that it was the successor to democracy, once the latter was doomed. From 1929 onwards Hitler's movement swallowed its rivals on the Right and turned the NSDAP into a coalition itself. The scale of its activity was always impressive; in rural South Hanover-Brunswick the Gau leadership held 1,200 meetings in the first quarter of 1930 alone. No one was left in doubt that the NSDAP intended to destroy the 'System'. Moreover, the skilful adherence to legality contrasted favourably with the frequently crude approach of rival organisations. In February 1932 the *Landbund* told peasants not to deliver food to the cities unless prices improved. Hitler then issued a statesman-like appeal telling farmers to bring out their produce as soon as possible as the country was short of foreign currency and could not afford domestic shortages.[38] The contrast between the two approaches was surely not lost on the land.

In terms of content NSDAP propaganda followed two broad principles; first, unremitting attacks on the Republic and all its works, linked with anti-semitism and anti-Marxism. The definition of 'communist' was wide enough to embrace virtually everyone against the NSDAP on the grounds that by opposing it they were contributing to the victory of Marxism directly or indirectly. Armed with this convenient logic, party speakers hammered home the fallibility of their opponents at every peasant meeting, whether it was a mass demonstration in the Hunsrück or whether it was at one of the thousands of village halls where they spoke. Sometimes not even the latter facility was available and the meeting was held in a local pub, or even in a field with a swastika over the hurriedly erected platform. In Westphalia, individual canvassing was the rule; the LGF, Meinberg, attended all rural co-operatives' or Agricultural Chambers' assemblies and handed out party membership forms or propaganda leaflets to the departing peasants. This turned out to be so successful that Darré recommended it to all LGFs.

Hand-in-hand with promises of a glowing future as contained in the March 1930 programme went the attacks on the 'System' as such: this was carried out at various levels, by speaking of financial distress caused by democracy, by attacking individual leaders as well as by playing on the bogeys of Marxism and anti-semitism, alleging that democracy was only really a sham and secretly controlled by Jewish wire-pullers. Full attention was always given to the first point; distress in the Bavarian Alps was said to be due to the flooding of Germany with foreign food and the lack of any adequate import protection, the latter being a sore point with the farmers.[39] When Dr Dietrich, then Minister of Agriculture, spoke favourably of a certain brand of imported wine, the NS press demanded 'What does the (German) wine-grower who voted for this party (DDP) think now?'[40] That the farmer's plight was caused by reparations was also a favourite theme in the appeal to nationalism on the land; Brüning's emergency programme in 1931 was hailed as 'New mass taxes for the German slaves of Young'. Commenting on an Agricultural Chamber report on farm distress in Württemberg the leader of the NS group in the Landtag blamed this latter fact on the Young policy which was bleeding Germany white.[41]

Linked with attacks on current governmental policy was the tactic of demanding measures which could not be granted at the time in order to paint the Republic's record even blacker. For example, in August 1931 following a meeting of the party's agrarian cadre the government was asked for immediate assistance comprising lower interest rates, a debt moratorium, lower prices for fertilisers and reductions in imported food.[42] All this was cleverly put together to achieve the maximum appeal to farmers' grievances at the time.

Personal assaults on political opponents was a regular feature of NSDAP propaganda with anti-semitism frequently employed as the basis. The party seemed convinced that this was efficacious as a propaganda line, and a directive was sent out to all functionaries in March 1931 stating that 'the natural hostility of the peasant to the Jews . . . must be stirred up to boiling-point'.[43] Other movements had been inciting peasants against the Jews long before the advent of the NSDAP in any case, notably the *Bund*; indeed, it had banned Jews from membership at its first ever general meeting in the 1890s, and had remained anti-semitic since. The NSDAP now extended this to include virtually all rival politicians or opposition leaders. This kind of propaganda has to be seen in conjunction with the claims made by the movement that it alone could save the farmer; destruction of faith in other parties and groups was an essential part of this operation.

Typical as an example of such tactics was the plan to destroy the faith of the peasants in the Stahlhelm by alleging that one of its more prominent leaders, Dusterberg, was a Jew. Similarly, a general instruction was issued to the agrarian cadre to pillory the von Papen cabinet as being controlled by wire-pullers in the form of Jews and Freemasons.[44]

In terms of political abuse the party was quite remarkably eclectic. Any pejorative description of an opponent sufficed. All attacks on the

DNVP to be made by the agrarian apparatus were ordered to be directed at Hugenberg the leader, as he was in trouble with the younger element of the party; he was to be labelled as 'reactionary'. Then in October 1931 Darré told his agitators to characterise the Brüning regime as being synonymous with Kerensky's, in that the Chancellor was moving to the Left and would be eventually forced to give way to pure Bolshevism.[45] Even personal failings of opponents were frequently alleged in order to create distrust among potential voters: in Schleswig-Holstein the chairman of an anti-NSDAP farm union was accused of having misappropriated official funds.[46]

If this type of attack seems naïve it must be pointed out that voters were not always capable of sophisticated discrimination; one farmers' meeting (not organised by the NSDAP) in Passau (Bavaria) demanded that all state expenses should be reduced to the level obtaining in 1913; on the same day farmers in Hanzenberg (Bavaria) demanded more state aid, and the extension of Osthilfe to the Bavarian Ostmark.[47] Party speeches were being directed at people who apparently saw nothing illogical in demanding less taxes and more state expenditure simultaneously.

Apart from the mass appeal to the agrarian community in general the party made great efforts to win over certain individual sections. Darré believed that women were often more active workers for the NSDAP on the land than their menfolk, so that the rural vote could hardly be brought in without them. Consequently he ordered his agitators to work hand-in-hand with the Agrarian Housewives League, a non-party organisation, wherever possible. Meetings were convened on occasion for women only, and were claimed to be highly effective.[48] The NS press also participated in this campaign drawing its readers' attention to the plight of the overworked rural housewife, placed in a hopeless financial position in recent years; only Adolf Hitler could relieve this misery.[49]

Equal attention was directed to another special group, those peasants too poor to be able to pay a party subscription to the NSDAP, but who were nevertheless potential voters. A special organisation was founded in September 1932, the *NS Bauernschaft.* Four months later all its branches were placed directly under the agrarian cadre to facilitate unity of control. Members paid only sixty Pfennigs monthly instead of the normal party fee of RM 1.50. Until such time as they could afford to enrol officially, they were indoctrinated inside the framework of the new organisation.[50]

Thus a wide variety of means was utilised on the land, either to persuade the rural population as an entity or in appeals to different sections of it. How all resources could be co-ordinated in one campaign can be seen in Darré's instructions to his cadre for the November 1932 elections: plans were laid down as though for a military assault, probably a reflection of Darré's service as an artillery officer in the war.[51] Section A, entitled 'Means', listed the various organs available, such as the *NS Landpost*, an agrarian magazine *Deutsche Agrarpolitik*, the normal party press and pamphlets and brochures. Section B delineated the use of all these different weapons as additions to all the usual party meetings. Beginning on 10 October there was to be a unified propaganda approach, described as 'advanced artillery fire'. This was to attack the entire cabinet, the main theme being its abandonment of agriculture to capitalist speculation. Obviously this was in line with the usual party approach anyway, as already indicated here. The difference was the concentrated nature of the campaign over a period of six days. Then use was to be made of a Strasser speech on interest-slavery, serving as an introduction to the second wave of the assault, centring around unemployment and rural indebtedness. The third stage would commence ten days later on 30 October, based on personal attacks on the DNVP leader Hugenberg, as the man behind von Papen. The dumping of foreign food was to be highlighted here.

All the threads in the weave were pulled cleverly together in one unified pattern, designed to exercise the maximum effect on those for whom it was constructed. Virtually every conceivable resentment on the land was catered for, particularly rising indebtedness, with anti-Semitism, anti-Marxism and anti-liberalism included. In effect attacks either on capitalism or on Communism were in themselves anti-semitic, as Jews were always designated as the leaders of both by the NSDAP. In retrospect the tone is naïve, but after all listeners were psychologically prepared by their grievances, sometimes exaggerated, sometimes genuine, against the existing government. The NSDAP was in sporting terms batting on a very good wicket. Democracy appeared unable to cope with the crisis in the eyes of the peasants, and indeed by its very nature.

Germany seemed to many to have become a veritable battleground of competing lobbies and pressure-groups, all noisily demanding financial succour from the state. All too often the pluralistic society was seen as being in itself the origin of such divisions, a belief on which the NSDAP played. Democracy was unacceptable for a large part of the German population particularly at a time of economic depression which seemed to call for a united society on Hitlerian lines, rather than for an internally-divided one.

These divisions were reflected in current party politics. Typical of these was the attitude of the Communist party to the SPD; instead of trying to form a common front it abused it almost as though it were the same as the NSDAP. Hence the satisfaction expressed in one NSDAP report from a rural area concerning a Communist pamphlet which had devoted four-fifths of its contents to attacking the SPD and only the remainder to the Hitler movement.[52] The peasantry viewed Left-Wing parties with the greatest possible suspicion in any case; much of its initial opposition to National Socialism arose from fears about the implications of the second word in the party's title. The Social Democrats had, in fact, adopted a pragmatic approach to the peasant holding at their Heidelberg Conference in 1925, declaring that as family labour only was involved it could not be designated as exploitation, but feeling on the land remained sceptical. One agrarian weekly even alleged that the SPD wanted rural poverty in order to drive the peasants to Socialism.[53]

The Communists had also taken a fresh look at the peasantry in 1925, and divided it into four classes, according to farm size.[54] The upper two were declared to be bourgeois and the lower proletarian: this analysis is clearly based on the assumption that socio-economic class determines political allegiance, which in itself is question-begging, to say the least. The party had founded its own farmers union, but the strength of only 10,000 is eloquent testimony for peasant rejection of Marxism.[55] Even if the parties of the Left had not been divided it seems unlikely that they would have found much response on the land. The Marxists own ideological approach was probably a factor here.

Except for opposition from the Catholic Centre party the Hitler movement had a clear field of action in rural areas. The NSDAP did have in general a less enthusiastic reception from Catholics than from others, in both rural and urban areas. Statistics in confirmation of this are available both for Lower Saxony and for Württemberg, at opposite ends of the country. Rural indebtedness however, was generally higher in Protestant regions than in mainly Catholic ones.[56] Any idea that the difference in voting preference was occasioned solely by religion would be a marked over-simplification. Additionally, the Catholics had their own political organ, the Centre Party, which held its share of the vote steadily around 15 percent up to 1933. The NSDAP advance in non-Catholic districts may have been primarily due to the existence there of a political vacuum on the Right as in the case of Schleswig-Holstein. In other words, religious belief was by no means the only factor in the situation, and may not even have been the main one. In any case, Hitler, Goebbels and Himmler were themselves originally of the Catholic faith.

Moreover, even if many of the Catholics did not vote for Hitler, associations sometimes said much the same sort of things as the NSDAP itself anyway. When the CBV attacked NSDAP proposals as damaging for agriculture in 1923 it concluded that 'like the National Socialists we also fight against the Jewish-materialistic spirit'.[57] Thus the ground was being prepared for the NSDAP to some extent, despite ostensible opposition.

In the final analysis the abrupt rallying of the peasants to Hitler's flag poses not so much the question 'why did it happen? ' but rather 'why didn't he appeal to the rural inhabitants sooner? '. After all, in *Mein Kampf* he had written that the path to success in mass elections lay in a concentration upon those classes whose livelihood was being threatened by modern developments. In the 1920s this applied quite clearly to the lower middle-class artisans, small shopkeepers and peasants far more than it did to the industrial workers. Under these circumstances Hitler's pre-1928 orientation towards the cities seems in retrospect to have been a clear failure in judgement. His desire to beat Marxism on its own ground apparently blinded him to the biggest reservoir of votes in the whole country, ready and waiting to be tapped by the right party. Only abject failure in the Ruhr re-directed his gaze. It has been pointed out that one facet of Hitler's character was his concentration on one thing at a time to the exclusion of all else. Here perhaps is another instance. Certainly his conduct as a commander of the besieging forces was curious between 1925 and 1928 after nearly three years of battering away at the strongest point of the Weimar citadel, the working-class, he suddenly discovered that the back gate had been yawning open all the time.

This is, of course, not to deny the obvious effects of the depression on Germany; after 1928 the party got far more votes in urban areas too than it had before. But with the advantage of hindsight it is possible to query Hitler's tactical skill as a party leader in respect of the general line pursued up to early 1927. Moreover, it was men such as Werner Willikens and his comrades in the agrarian apparatus or the local political leadership travelling tirelessly from one village to another month after month to bring the party's message to the peasantry who won the rural vote anyway. The real strength of the party machine as far as the country areas were concerned lay precisely in its local branches rather than in far-off Munich.

This impression is heightened when the very number of such cells is considered. As early as 1930 a local government official in the north reported that virtually two-thirds of all the villages in his area each had its own NSDAP branch organisation already,[58] and this at a time when the population was by no means always acquainted with the top leadership of the party. Police reports for Schleswig-Holstein in 1929-30 frequently misspelt Hitler's own name. These remarks are not intended to advance the theory that Hitler was not an orator or leader of frequently astounding skill, but simply to put the success of his movement into proper perspective, at least as far as the agrarian community was concerned. A proper balance has to be struck between Hitler's contribution and those of such subordinates who held seventy meetings in one rural district in one month alone.[59] Of course, their leader often spoke to mass peasants' demonstrations, but it seems very likely that the local cell was the biggest single factor in vote-catching. It is important to remember here that few peasants had radios and tended to rely on local events for their information.[60] Consequently for all Hitler's personal talent he was probably little more than a name to most peasants in northern and eastern Germany prior to 1933. As the undermining of the *Landbund* and the Agricultural Chambers had also in the final analysis to be carried out locally we should be careful not to over-stress Hitler's own role in winning over the country people and their associations. It was an operation where the *NCOs* were probably at least as important as the Commander. In any event, the part played by the peasantry in the north and east especially in elevating Hitler cannot seriously be denied.

4

THE NSDAP's CO-ORDINATION OF AGRICULTURAL ADMINISTRATION AND THE ACTUAL LEGISLATION IN 1933

The discontent and confusion on the agrarian scene during the final stages of the Republic were doubly significant. First, as an explanation of what Konrad Heiden called 'Hitler's meteoric rise' and secondly as an important factor in facilitating the erection of the centralised administrative framework which the party clamped so firmly and above all so rapidly on German agriculture. A feeling among farmers that it was time for unity to replace the divisions of the past undoubtedly played a large part in this. In addition, the NSDAP skilfully exploited the farmer's betrayal complex. The party was now at pains to underline how much times had now changed for the better for agriculture.

Hitler's accession was greeted enthusiastically in many quarters on the land. Von Kalckreuth was the first to get in his congratulations, the day after Hitler's appointment. On the same day the Agricultural Council passed a resolution greeting the new government. These were only the first in a positive flood of welcoming messages from all parts of the Reich expressing the confidence of farm associations in the new régime. In March a laudatory *Landbund* national resolution referred to the Führer's 'work of salvation', to which it was proud to have contributed.[1]

Hitler skilfully used this fund of goodwill, particularly by promising better times ahead; in his election address he stated that the government would relieve the German peasants' misery within four years. On 23 March in the Reichstag the Führer declared that profitability must be restored to agriculture, in effect at the consumer's expense. However hard for the latter this would be, it would be preferable to the ruin of the peasantry. Here was a clear offer of better prices even if the public had to pay more: Hitler rammed his advantage home on May Day with a declaration that national recovery had to start with the peasants, and that future policy would be the opposite of the treatment received from previous governments by agriculture. His sentiments did not go unheeded; a laudatory speech to the Agricultural Council on 5 April was reprinted in part in a Catholic farm journal under the heading 'The honouring of German peasantry by Adolf Hitler'. The newspaper opined that the speech should be put up in public offices all over the country.[2] This was a far cry from the comments which the same source had made about previous governments.

But among agrarian leaders some remained whose presence the party found unacceptable; in March 1933 Dr Hermes of the Christian Farmers Union was arrested on a charge of having misappropriated funds, and given four months imprisonment; this effectively removed him at the time of co-ordination, which he might have opposed. The Westphalian branch of the CBV, long in favour of unity of professional representation for agriculture, proposed new discussions, which took place on 27 March, the unfortunate Dr Hermes not being mentioned. A new leader, von Lüninck from the Rhineland, was chosen, and it was agreed to approach other unions for further discussions on unity.[3]

Meanwhile the *Landbund* had been dealing with the same topic; Darré knew of this in advance and ordered his rank and file to attend in order to ensure that the union came to the right decision from the NSDAP's standpoint. In this they were successful, and on 22 March the *Landbund* expressed the belief that only through unity would agriculture be reconstructed as a pillar of the national state.[4] The basis for this had now been created, and on 4 April a press notice appeared announcing that a joint meeting of the agrarian apparatus of the NSDAP, the Christian Farmers Union and the *Landbund* would take place at the invitation of the latter's management committee, which included both Willikens and von Sybel.

There von Kalckreuth proposed that a unified organisation be erected by 1 January 1934: under the title *Community of the German Farming Profession* a provisional body was formed at once with Darré as chairman. In principle, this had four representatives from each member association, but Willikens was on as a *Landbund* representative, so that the NSDAP had five members out of twelve. However, in May

Kalckreuth was arrested on the now-familiar charge of corruption. In his enforced absence Meinberg, LGF in Westphalia, was elected as a President of the *Landbund* and took Kalckreuth's place on the *Community* Committee. The balance of representatives was now even, but Dr Hundhammer of the Christian Farmers was arrested in June and sent to Dachau. The NSDAP now held the majority. In the same month its control was made absolute by the appointment by Darré of Meinberg as deputy chairman: he simultaneously announced that the *Community* would be divided into four departments, all of which were given National Socialists as leaders. Existing farm unions could either affiliate with it, or go into voluntary liquidation: the *Landbund*, the *NS Bauernschaft* and the League of Agrarian Housewives took the former course; the Christian Farmers Union (founded 1862) wound itself up in September:[5] the *Deutsche Bauernschaft* and other smaller unions did likewise. The leader of the *Deutsche Bauernschaft,* Heinrich Lübke, had in any case been consigned to a concentration camp in the previous April. In the space of a few months the whole picture of agrarian organisation had been transformed; the historic (and mutually antagonistic) unions had given place at national level to one comprehensive corporation under Darré's control.

By April he was issuing guidelines as to how the regional co-ordination of unions should take place; the LGFs must get on the governing committees by election or 'some other way'. Darré was concerned about the possibility of deception by branches not controlled by the NSDAP, in that such a body might enrol its governing committee in the party and then announce that it had NSDAP leadership. To obviate this he laid down his own criterion of a National Socialist, which meant someone actually approved by the relevant LGF as a genuine party member.[6] Thus, Darré guarded the new corporation against dangers inherent in the sudden onrush of new members to the NSDAP who would not really be genuine supporters.

In fact, takeover in some regions had preceded this directive: in Schleswig-Holstein a farm union which criticised the government's election manifesto was attacked by the SA. On 1 April the police occupied its offices on the pretext that state funds had been misused. The *Kreisleiter* of Flensburg declared that only the union already affiliated to the *Bund* could be accepted. On the 5 May Struve, the regional LGF, declared it to be his duty to unify unions, chambers and co-ops: the union already attacked was then liquidated together with all others not yet affiliated. The latter was left as the sole representative body, and new elections gave its management a NSDAP majority.[7] Meinberg acted similarly in Westphalia; a corporate union like the national *Community* was established at regional level under his chairmanship, with the same Christian Farmers, NSDAP cadre and RLB tripartite membership on the council. Thus, in the regions co-ordination under NSDAP control was pushed through as efficaciously as at national level.

The semi-official Agricultural Chambers were equally necessary and so a campaign to consolidate NSDAP power (where its participation in committee elections had already given it a foothold) began almost immediately. The SA began to arrest the Chambers' leaders who were in the DNVP. This move naturally aroused Dr Hugenberg, its leader and Minister of Agriculture in the Hitler government, to protest. In April he raised the matter in Cabinet and got a very strong reaction from Goering who after some exchanges rejoined that until new elections were arranged he could not hold the SA back.[8]

At regional level similar tactics were adopted; in Württemberg the NSDAP took control first of the government by demanding two ministerial posts out of three, although in the elections they had polled slightly under half the votes cast in alliance with the DNVP. There was little possibility of opposition. A National Socialist was installed as regional Commissioner and on 8 March named the local SA leader as head of all local police forces on the pretext that public order was in danger. The leader of the party in the *Landtag,* Mergenthaler, became head of the administration. He soon dealt with the local Chambers by a new law transferring all duties of their general assembly to a committee of management, which would be named by the government: the voting rights of the members of the assembly was terminated. Again it was alledged that the Chambers no longer corresponded to the realities of the situation.[9]

In Schleswig-Holstein, a slightly different approach was adopted; Gauleiter Lohse appointed a special Commissioner to take over the local Chamber in April. The existing chairman had already been told that if the present management committee, which the NSDAP did not control, did not resign its members would be arrested: at a meeting on 8 April the hint was taken, and the committee stepped down in favour of the Commissioner.[10]

The 40,000 rural co-operatives were an important feature of the

German agricultural scene. It was regarded as essential to bring them under NSDAP control, and in February a plan was devised by Arnold Trumpf, a member of Darré's office.[11] Current leadership was estimated to be mainly in the hands of the DNVP and the Centre Party. Similar infiltration tactics to those employed against the *Landbund* were to be used against it. Attendance at all meetings was to be a duty for party members: they must agree among themselves beforehand about candidates for vacant places on the management committees and then vote en bloc for their chosen man: this strategy must work 'as most co-operative meetings are poorly attended'.

In April two positions on the Co-operative National Council fell vacant; Darré intimated that he wished to fill these with Trumpf and Granzow (LGF in Mecklenburg).[12] As the election would take place in six days there was no time to be lost; Darré enclosed a list of the names and addresses of all existing councillors to be approached by the appropriate LGF demanding support for the NSDAP candidates: hints were to be dropped that non-compliance would entail a lack of future support for co-operatives by the NSDAP. In addition, telegrams were to be sent to the Council as a body by the LGFs, and articles were to appear immediately in the press stressing the necessity of incorporating the co-operatives inside the new unified *Community* structure. As a final effort Darré, accompanied in force by Willikens and Meinberg among others, entered the Council meeting on 20 April and demanded the election of the NSDAP to the management committee. The pressure was successful and three party members were elected to the committee.[13] Infiltration by the NSDAP as outlined in Trumpf's original plan of conquest was a contributory factor.[14] The following day the co-operatives were formally affiliated to the new *Community;* the department responsible for them was headed by Trumpf.

Again regional co-ordination paralleled the national; in Westphalia the local general council held a meeting in June and elected Meinberg and three other members of the agrarian apparatus to its ranks. Von Lüninck was elected for another vacant place, as he was by now committed to support of the new régime.[15]

Other quite important bodies brought under NSDAP control included the League of Rural Trades *(Landhandelsbund)* of which Darré became head in April. The farmworkers unions participated in the fate of workers unions in general in the Third Reich. On 2 May there was a general arrest of leaders, on the charge of embezzlement of funds, as many as 500 being taken into custody from the Landworkers Union alone. From now on only Dr Ley's Labour Front was permitted as the organisation for working-class representation. Finally there was the Agricultural Council, semi-official and highly influential: to this Hitler had paid court in April by attending its general meeting, to which he addressed a eulogy on the agrarian population. He referred to the unity it was now finding and assured his hearers that the Cabinet was ready to take such measures as would ensure the future prosperity of the agrarian sector.[16] His remarks about unity evidently sank home; on 12 May the Council held a meeting at which the chairman accepted the government viewpoint on this matter and consequently recommended that Darré be elected as his successor.[17]

Additionally the party's agrarian office infiltrated its representatives into positions of influence in the state machinery in order to ensure that the 'correct' decisions were taken, for example, in Bavaria. Like all states except Prussia and Mecklenburg it had had no separate Ministry of Agriculture prior to the NSDAP accession, rural affairs being administered by the Ministry of the Interior. The official responsible for this department was in fact the LLF for Bavaria, and a party member. He issued a ruling that in future LGFs would be permitted to make proposals direct to the Ministry of the Interior on agriculture. Similarly, before the Ministry made any decision on its own ideas it would consult the LGFs: wherever the Ministry wished to deviate from the latter's advice, the agricultural department would have the final word: this department was under him anyway, so that the LGFs could, in effect, use him to work their will on the state apparatus. Darré was so struck by this system that he recommended it as a model for all LGFs.[18]

By autumn 1933 the NSDAP controlled all agrarian organisations and had won influence over local state policy as well. When it is recalled that in September 1930 the party had not even possessed any agrarian apparatus at all, the momentum of the gains made seems astounding. The rural community had been taken by a mixture of propaganda, and skilful infiltration. On the importance of the latter tactic there can be no doubt. As Darré himself said unification could never have taken place without the work of Willikens in his capacity as president of the *Landbund.*[19] This applied equally at local level; in March 1933 the *Bund* branch chairman in Alfeld (Lower Saxony) was Herbert Backe, one of the four NSDAP members later made departmental heads in the *Community.*

But despite these points it remains highly improbable that such a rapid takeover could have occurred had the victims been less willing. The words used by the then French Ambassador to describe the NSDAP takeover of the country in general can well be applied to the agrarian scene in particular: 'The astounding things in this revolution are the speed . . . but also the facility with which it is everywhere completed, the scant resistance which it encounters'.[20] However, it must be said that in many quarters there was a fundamental misunderstanding of the extent of the changes which were taking place. The fact that there were only three members of the NSDAP in the Cabinet initially was doubtless significant. When the government in its election manifesto of February promised to relieve rural misery in four years, one farm journal solemnly informed its readers that it would never cease to remind the politicians concerned of these words, apparently under the impression that comment would still remain free. In October of the same year it printed its last issue.[21]

Furthermore, the NSDAP co-ordination took place in the context of a legislative programme which was beneficial to agriculture. When Hitler visited the Agrarian Council meeting in April, for example, the chairman actually thanked him for the policy that the 'government of national uplift' had so far pursued. Initially the NSDAP was fulfilling its solemn pledges to aid agriculture and understanding of its acceptance on the land can hardly be grasped without taking cognisance of this fact, which entails an examination of the legislation concerned.

Whilst Darré at one level had been unifying agriculture Dr Hugenberg, as Minister of Agriculture, had been grappling with the agrarian problems inherited from Weimar; these centred around three main points, namely, falling prices for farm products, a consequent rise in indebtedness, and an imbalance of production typified by a continuing grain surplus in a land importing 60 percent of its animal fats consumption.

Hugenberg himself believed that the chief fault of previous administrations was that agriculture had never been dealt with as a unity in policy terms. He now proposed to settle its problems by a series of simultaneous measures; these included the compulsory mixing of butter and margarine to use up surpluses of the first, plus the reduction of imported feeding-stuffs, of which Germany had bought some two million tons abroad in 1932. This latter point was linked with the grain surplus, as pointed out by von Rohr (DNVP), whom Hugenberg had brought in as his Secretary of State: duties on fodder imports would pay for official price support of domestic grain, the surplus of which could be used in lieu of foreign maize and barley as feeding-stuffs by German farmers. As for farm incomes, Hugenberg believed that the only solution was to cut Germany off from the world market prices which Hitler accepted.[22] This was simply a continuation of Weimar protectionism. So heavier duties on livestock and meat were introduced, while those on certain fats were quadrupled; grain prices were supported and those for flax improved.[23] Von Rohr explained the reasoning behind the government's tariff policy in a radio speech on 22 February. As far as fodder was concerned, the Cabinet accepted Hugenberg's proposals and in June 1933 the maize monopoly organisation instituted in March 1930 was replaced by a body controlling all grains and fodder, plus oil-cake, etc.

The disposal of the grain surplus and the cutting-off of Germany from world meat and livestock prices had not proved too difficult, but fats were another thing entirely, as Schleicher had found to his cost. Hugenberg defended his proposal for implementing the compulsory mixing of butter and margarine at a series of Cabinet meetings, as part of a general 'Fats Plan' to restore profitability to the dairy farmers. He also wished to reduce margarine production by law, and tax it heavily, to encourage the use of home-produced butter. The proposal for mixing proved to be the greatest obstacle, since Hitler himself was reluctant to embark on this policy, which the NSDAP had never supported. He declared in Cabinet that the price increase would be unacceptable to the poor. Hitler also feared that rationing might become necessary under such a scheme. The Führer's objections were taken up by Frick, the Minister of the Interior, who foresaw possible political repercussions in the shape of unrest among the population if price increases were established.

To this Hugenberg could only reply that prices were falling and something must be done; he felt that if dairy-farming were not made more profitable farmers would switch to producing grain of which the country already had too much. Hitler asked, for the second time, for the debate to be adjourned: he said that he did not like the proposals but that he felt himself obliged to help the peasants and wanted time to think. On 18 March Hugenberg returned to the attack with a memorandum to the Cabinet claiming a causal connection between dairy products' prices and rural indebtedness which no doubt was justified. To deal with Hitler's objections about the poor he proposed a levy on margarine manufacturers, in order to raise RM 190 million annually, to be used as a subsidy.[24]

The result was a compromise; the new 'Fats Plan' taxed margarine at 50 Pfennigs per kilo, which was used to subsidise a coupon system enabling the poor to buy at reduced prices: production of margarine was limited to one-half of the previous year's output by a series of quarterly decrees: the price of butter and milk was raised to the domestic producer: surpluses were bought up and stored by a new governmental agency for controlling fats imports. There is no reason to doubt the claims made that the plan was a success. By June the domestic prices for butter and lard had gone up on average by one-quarter, whereas in the corresponding three months of 1932 both had actually fallen.[25]

The 'Fats Plan' is interesting for two reasons. Despite Hitler's initial objections he had to accept price increases for food, none the less, which bore on the public. In June 1933 the average consumer was devoting nearly a quarter more of his general food outlay to fats than he had done five months previously. This, moreover, was at a time of general financial distress. Hitler's Reichstag speech of 23 March had thus turned out to be an accurate forecast. A degree of profitability had been returned to agriculture, and it had been at the consumer's expense. The agrarian price-index rose by 5 percent in the first half of the year for farm produce in general.[26] This may sound slight, but it was after all the first time for three years that it had not actually fallen. This makes rural support for the new régime readily comprehensible.

Hitler had taken a calculated risk in his policy of furthering agriculture, the effects of which could well have been disastrous for him in view of the increased cost of living. How well aware of this he was was made very obvious in his demand to be furnished with monthly reports in future regarding the fats situation 'in appreciation of the political dangers'.[27] The Führer always had a lively sense of the risks for the NSDAP contained either in food shortages or rising prices. Their effect on wartime morale, and their contribution to German defeat and the ensuing political unrest was apparently a traumatic experience for the people of Hitler's generation, especially on the Right. Thoughts of popular disturbances seem never to have been far from his mind, at least until 1935.

Secondly, the NSDAP had departed in one sense from its pre-1933 propaganda line. It had always been hammered into the peasants that they could not expect to be saved at the expense of other sections of the community, for example, in the March 1930 programme. In the presidential election campaign in April 1932 exactly the same sentiment had been expressed by Hitler himself. Indeed, the farm unions had always been pilloried by the NSDAP precisely because they had demanded one-sided assistance from the Weimar governments. Now in office the party seemed to be contradicting itself. Partly this was doubtless an example of Hitler's well-known flexibility and pragmatism over tactics. But partly it may well have been a reflection of the influence of Hugenberg and von Rohr of the DNVP. Whilst the Führer consolidated his internal power, economic policy had been left largely in other hands, in the coalition cabinet.

This applied equally to debt-relief. Here again Hugenberg had his own viewpoint, which clashed with Darré's. Higher prices would help fight future indebtedness but in the meantime something had to be done about the current situation, since debts in 1931-32 had cost agriculture over RM 1,000 million in interest payments.[28] Hugenberg's own party had proposed a general debt-relief scheme in 1931, on which was based the new legislation in June 1933.[29] All farmers now unable to meet their obligations were to register this by 30 June 1934: recipients of Osthilfe were excluded. Debts taken over by the state were to be subject to variable interest rates, depending on the level and their nature.

Hugenberg initially met with stiff opposition from Darré. On 11 May the latter, with staff from the party agrarian office, met Hugenberg and von Rohr from the Ministry in what was in effect a clash of ideologies. This much had already been made clear by Darré, who had learned of Hugenberg's draft in advance following which he expressed the view to his staff that one could not have approached the question in a more capitalistic way than the Minister had done. Three days previously he had complained in a letter to Hugenberg himself that the whole draft had been based upon purely economic considerations. The new proposals offered debt-relief only to those farms deemed capable of financial reconstruction. To Darré this was unacceptable on racial grounds, as many peasants might have valuable blood to contribute to the race but perhaps would not qualify for the new scheme economically. He referred to Hitler's expressed desire to preserve the peasantry as the life-source of the nation.[30]

On 11 May even the interest rates caused dissension, as Darré wanted 3 percent for all debts. The Minister drew his attention to the economic situation as precluding so low a figure and suggested that in the long run one Pfennig extra per litre on the milk price would help the peasantry more than Darré's desired cut. The question of non-viable farms does not seem to have been raised at the meeting, at which the Agrarian Office came out second-best, in the sense that its objections seem not to have altered the draft proposals.

In Cabinet, however, Hugenberg encountered opposition of another kind to the interest rates, especially from the Finance Minister, Schwerin von Krosigk, who felt them to be too low. Dr Gürtner, the Minister of Justice, also recorded his opinion that the government was being asked to pay too much. One source, in fact, estimated the total cost to the state as RM 460 million.[31] Eventually, however, the draft was accepted.[32] The Minister of Agriculture had thus piloted the measure between the Scylla of Darré's racialism and anti-capitalism on the one hand, and the Charybdis of more orthodox financial views on the other. However much Darré wished to assist the peasants because of their 'valuable blood', the fact remained that Germany's financial position was weak; the Budget deficit was currently running at RM 945 million per annum. Thus, the Hugenberg Act as passed was a compromise to a certain extent, but one which leaned rather more towards financial orthodoxy than towards National Socialist racialism. The law was ultimately successful, as approximately 10 percent of all farmers were substantially relieved of their financial burdens and given a breathing-space to rebuild their economic position.[33]

Side by side went the extension of the protection against foreclosure initially introduced under the Osthilfe plan. The existing arrangements were due to lapse in June 1933, and Frick asked for an extension in February. Dr Gürtner believed it should apply to public debts only rather than for financial obligations in general, whereupon Hitler pointed to the necessity of satisfying the peasants, presumably politically, which was also Frick's motive.[34] Eventually foreclosure was forbidden before December 1933. Here political demands by the NSDAP had defeated the financial orthodoxy of non-party ministers such as Dr Gürtner.

Hugenberg's term of office was a substantial success; he dealt competently with outstanding problems, whilst the foreclosure protection measures of the NSDAP complemented his own debt-relief scheme.

Protection against imports had been extended, prices improved. To a certain extent he had been given a free hand at a time when the NSDAP was intent upon consolidating its political power-base in the country.

However, in April Darré telegraphed to Hitler announcing 'strong unrest' among the peasants over the secrecy under which the new debt-relief law was being prepared; he underlined his personal opposition by enclosing a copy of the letter he had sent to Hugenberg with his telegram. Inside his own cadre Darré had already begun the work of undermining the Minister's position with a directive telling his LGFs not to attack Hugenberg or von Rohr by name but to speak rather of the 'capitalistic agrarian front' so that everyone would know who was meant.[35]

By late April his hostility was apparently being taken up in Cabinet. Frick asked Hugenberg if Darré was being kept in touch with proposed agricultural legislation, as he had not been acquainted with the details of the debt-relief Bill. Hugenberg replied that he had invited Darré to a meeting to discuss it but that he had not turned up.[36] The original telegram to Hitler may have played some part in setting off this pressure on the Minister of Agriculture, although Hitler now wished to amalgamate the NSDAP and the DNVP, which may also have been a factor in the situation.[37] Certainly pressure on the latter now intensified; by the end of May Hitler was threatening to disband its youth organisation. On 11 June some DNVP members went over to the NSDAP and the NSDAP press began to speak openly of the 'collapse of the German Nationalist Front'. Ten days later all DNVP subsidiary groups were formally disbanded by law, on the grounds that they had been infiltrated by Communists and Social Democrats, which seems rather unlikely. Increasing subversion of Hugenberg's own position came in May, when a peasant meeting in Thüringen demanded his replacement by Darré (after a speech by the latter). A similar request came from East Prussia.[38]

Hugenberg might have survived the pressure had he not precipitated a crisis over his own personal qualities as a result of the London Economic Conference, at which he was a delegate. During the course of it he presented a memorandum, not apparently approved by his fellow-delegates, on the question of Germany's debts, which he said could only be met if the country's colonies were returned. This caused rumblings in the Franco-British press about the 'imperialistic goals' of German foreign policy: this was embarrassing to Hitler who was currently engaged in a campaign to persuade European powers of his peaceful intentions. When delegates returned to Germany on 23 June to render a progress report, Hugenberg found himself isolated in the Cabinet. First he tried to put the blame for the indiscretion onto Dr Possee, a senior Civil Servant in his Ministry, and asked that he be dismissed, a suggestion which met with no approval at all. Hitler remarked that Possee would loyally follow 'if he receives an exact march-route': the implication appeared to be that up to now he had not been fortunate enough to get one from Hugenberg.[39]

Four days later the Führer announced to the Cabinet that Hugenberg had resigned, citing this incident as grounds. He also revealed that he had previously asked the DNVP leader to accept Herbert Backe of the NSDAP as Secretary of State in the Ministry of Agriculture prior to the resignation. Hugenberg was also said to have agreed that it would be best for the DNVP to disappear completely. Hitler did not give the full facts behind this latter point; in fact, Hugenberg had been making desperate efforts behind the scenes to keep his party intact including an attempt to procure an interview with Hindenberg for one of the party's leaders, an ex-chairman. Only when that fell through did he then accept the inevitable.[40]

From the foregoing it is clear that Hitler intended to destroy the DNVP, although from the alleged request to Hugenberg about possible amalgamation made about the end of April he may originally have been willing to accept that as an alternative. Once Hugenberg had rejected the idea, he certainly sealed his party's fate.

Equally it seems pretty clear that Darré always wanted to get rid of him as a person, and set out systematically to undermine his position among the peasants and to destroy Hitler's confidence in him. In January Darré had written to his subordinates informing them that Hitler had promised him the post of Minister of Agriculture.[41] Darré would surely never have put this statement in writing had it not been true; presumably Hitler eventually chose Hugenberg for two reasons, firstly, to give an air of national coalition to his administration, and secondly, he may have wished to afford Darré the chance of concentrating wholly on the co-ordination of agriculture, in which case Hugenberg's dismissal was only a question of time.[42] Some support is lent to this theory by the post-war interrogation of Werner Willikens, who declared that Hitler had given him a verbal commission 'to lift Hugenberg out of the saddle'. When Willikens turned down the demand Hitler took it amiss, as a result of which he lost the post of deputy to Darré, which went to Meinberg.[43]

This has the ring of truth about it, since clearly some explanation has to be sought for the passing-over of Willikens in favour of the Westphalian LGF. He had been in the party long before Meinberg, and had also been official spokesman on agricultural matters in the Reichstag since 1928, and was author of the semi-official book on agrarian policy *NS Agrarpolitik.* He had been given priority for a place on the *Bund*'s governing body, being the official NSDAP candidate rather than Meinberg in December 1931. Finally, until April 1933 Meinberg's only party position was LGF in Westphalia, whereas Willikens was LLF for the whole of Prussia and therefore Meinberg's superior in the hierarchy of the Agrarian Office. It certainly seems curious that with such qualifications for office Willikens should have been offered nothing more exciting than the post of Agriculture Minister in Prussia and eventually Secretary of State in the REM when von Rohr resigned. This lends credence to his own account of the affair.

But there also seems a good deal which may be said on the other side; Hitler after all had used plenty of other politicians who were not in the NSDAP, so that Hugenberg's non-membership would appear not to have been grounds for his removal. The latter's own version of his interview with the Führer on 27 June supports this view: according to Hugenberg, Hitler had begged him emotionally to remain in office:[44] certainly when the Führer announced the DNVP leader's resignation he told the Cabinet that he tried to persuade him to stay. The evidence is conflicting when these facts are taken into consideration with Darré's January letter and Willikens' remarks. Perhaps they can only be reconciled by suggesting that Hitler wanted Hugenberg away from the REM in order to give Darré the post, whilst leaving the DNVP leader as Minister of Economic Affairs, which he had previously combined with direction of the REM. There is some evidence for this suggestion, as it has been stated that Hitler asked the DNVP leader to surrender the Economics Ministry only.[45] At any rate, on 29 June Darré became Minister of Agriculture as well as continuing in his previous posts. The vocational, state and party leadership of agriculture was in the hands of one man, co-ordination incarnate.

5

THE UNDERLYING PRINCIPLES OF THE PEASANT POLICY ADOPTED:

The War on Speculation

The government had now to a very real extent room for manoeuvre, as the cumulative effect of Weimar protectionism and Hugenberg's policies had been to cut German agriculture off from the world market, to which the lifting of reparations had also contributed. National Socialism had a clear field for the implementation of its most cherished principles. The point of departure was that the agrarian community was essential to the whole people and had to be furthered in consequence, but at the same time it had to serve society as a whole more effectively. This delineates the frontier with the DNVP policy, as demonstrated by Hugenberg, of just helping agriculture for its own sake. From July 1933 onwards, all aid was tied very clearly to corresponding responsibilities and duties for the rural community. As Darré said to his LGFs at Weimar in February 1935, it must be made clear to the peasant that we have not created special laws for him for the sake of his bonny blue eyes (*schöne Augen*) but because he has a job to do for Germany.

To ensure that he could fulfil this task he had to be sheltered from the play of free market forces, which if unrestrained might ruin him. Indebtedness in the agrarian community had, it seemed to the party, been produced in two ways. The farmers had been fobbed off with low prices in order to assist industry and exports. Additionally, the inheritance laws needed to be reformed. The NSDAP's answer to the first of these two problems was a completely new corporation (*Reichsnährstand*) which grew out of the 'agrarian community' established in spring 1933, based on reorganisation of marketing, and above all, on fixed prices for foodstuffs produced by the farmer. The second was resolved by the Law of Hereditary Entailment (*Erbhofgesetz*). As these two measures were complementary a brief general examination of their underlying principles will help to explain the nature, and internal coherence, of NSDAP agrarian policy.

The Agrarian Office had been working for some time on how to plan the reorganisation of food production and distribution; it had a special section called 'Marketing Organisation' at least as early as October 1931. Dr Reischle was mainly responsible for the actual details and by September 1932 was in a position to give a series of lectures on marketing arrangements in the Third Reich: that these had not yet been finalised can be deduced from a Darré directive warning speakers not to go into precise details about it in their speeches as the party itself was not yet fully clear on the subject. Indeed some members had a hankering after state control, including Gregor Strasser, but after a conversation with Dr Reischle at party headquarters in summer 1932 he was apparently won over to the Agrarian Office viewpoint:[1] this was based on the concept eventually hammered out by Reischle of an autonomous agricultural corporation rather than outright state control. In other words, Darré's Cadre had managed to impose its own ideas on the party. As the Strasser incident shows, not all National Socialists were initially of the same mind regarding agricultural administration, which had important consequences later.

A key concept in the thinking of Reischle and Darré was fixed prices for products as distinct from 'liberal materialism', under which 'all prayed to one God called the free play of the market forces'.[2] Prices had been a plaything of interests superior to the farmer, and had so denied him security. The peasant would now get a fair fixed price, in itself an incentive towards steady production. The party was convinced that the world economic crisis proved that in the long run the free market system must fail anyway. As one member of the Agrarian Office put it 'the liberal economic motor has an empty tank'.[3]

It does not, of course, follow that the NSDAP objected to liberalism on economic grounds only. As always, racial considerations took priority for the party. The people (*Volk*) was a living organism, grounded on individuals with varying hereditary gifts, but all inter-related in an organic community which operated according to natural laws. The liberal philosophy was essentially a negation of this concept, based as it was on the 'people' (i.e. any one nation) merely as being one stage on the way to a political system embracing the whole of humanity, an idea which denied the reality of inequality. Hence the NSDAP contrast between their concept *Volk* and liberalism, which by permitting economic selfishness (and miscegenation) reduced the population of any country to *Bevölkerung*, a mere statistical collection of disparate individuals. The French Revolution with its slogans of liberty and equality was seen as the origin of this downward trend, which explains the frequency of attacks launched in the Third Reich on the consequences of the year 1789. For Darré it was exactly the Revolution which had produced the modern egoism which now separated the individual from his family and from the fellow members of his own vocation/social estate (*Stand*).[4]

The party had strong practical grounds for wishing to preserve the peasantry as such. Here this term has to be employed other than simply as a synonym for the agrarian community. Darré distinguished clearly between peasant (*Bauer*) and any other food-producer in general (*Landwirt*). The first put service to community and family above material gain, the latter was engaged in agriculture merely for the income obtainable from it. It followed that only with the development of a monetary economy does the second type even appear, so that the real history of the German people is that of the *Bauer*; he talks of 'we', the *Landwirt* can only say 'I'. So for Darré, Klaus Heim was a real *Bauer* who fought to defend the family farm, in contrast to the 'bacon tariff patriots' of the other political parties.[5] Of course, Germany needed the *Landwirt* as he was a food-producer too, but he never attained any ideological status.

By June 1933 the scheme to reorganise marketing had been sufficiently worked out to enable Darré to give out the main guidelines in the *NS Landpost*. These envisaged a corporate body embracing all aspects of food production and distribution with a General Staff Office to plan ahead and co-ordinate in general. As soon as Darré was appointed Minister he hurriedly produced a law to authorise the new structure. He seems to have been worried that if this preliminary step were not taken on behalf of the Reich, individual states would start to reorganise agriculture on their own: the new provisional law therefore reserved any such powers to the central government alone.[6]

The bumper grain harvest which all Europe enjoyed in 1933 also compelled Darré to speed up the new measures, in order to avoid huge public expenditure on grain support, if a surplus produced low prices. Basically he wished to stagger prices, and thereby dissuade farmers from bringing all their produce on the market at once. The new organisation which the Minister foresaw would guarantee security in terms of income and production for the peasant, and also a steady supply for the public.[7] In addition, storage costs would fall on the individual producer and not on the public purse, since if prices were higher in winter than in autumn, the farmer would naturally hold back some grain to get the benefit. To make certain that he did not keep it all back, however, he would be compelled to surrender fixed percentages of the crop when they were needed in autumn. The new regulations were designed, in short, not merely to assist the producer, but to rationalise distribution and supply for the whole community.

The presence of six million unemployed in Germany was yet another factor to be considered. Public purchasing-power was limited, which lowered the current demand for bread. With a bumper harvest the amount of surplus grain would clearly have been considerable, and storage cost to the public correspondingly high.

September saw the publication of more specific legislation defining and describing the new organisational structure envisaged.[8] The original law had had first to be piloted through the Cabinet, where it met some opposition from two non-NSDAP members, von Rohr and von Krosigk, the Minister of Finance. Von Rohr in particular wished to control production by legal limitations on grain acreage rather than by using the fixed price system, under which output could be regulated by giving better prices for commodities most needed than for others. Darré rejoined that legal measures for controlling acreage would have entailed a vast bureaucracy. Hitler then spoke up for the proposed measure; he was of the opinion that control based on prices was more effective, since if farmers received a guaranteed price only for that quntity of food which the nation needed surpluses would be eliminated in future. His advocacy, and that of Goering, carried the day.[9] Thus, the NSDAP were successful in eliminating speculation in wholesale food prices; both the advantages and disadvantages inherent in the law of supply and demand were to be banished from the agrarian scene, of which more later. Darré had sold his ideas to the party.

The new body, the *Reichnährstand* (henceforth RNS), encompassed all food production and distribution, and included peasants, farmers, dealers, co-operatives, processing industries such as mills and dairies, wholesalers and retailers. Statistically it was vast comprising over 42,000 co-operatives, nearly 300,000 workers in the processing industries and another 500,000 in the retail trade apart from food-producers. The new Corporation was authorised to regulate production, marketing and prices where necessary which the RNS could delegate to individual sections if it desired. The powers given were both interventionary and supervisory, and no indemnity could be claimed for hardship resulting in individual cases from the new organisation. As a public corporation the RNS could fine members up to RM 100,000 or inflict prison sentences, or both, for transgressions. The parent body could close a retail or processing business both permanently or temporarily as it saw fit or refuse licences to trade to new firms. Membership of the RNS was obligatory; any trader in doubt as to whether his business was affected or not was advised to join to be on the safe side, as there were penalties for non-entry for those firms to which the regulations definitely did apply.[10]

All food-producers had to enrol naturally; when one *Landbund* member in Silesia attempted to resign from the union on the grounds that he did not wish to participate in any unitarian structure the local LGF announced that he would regard any such attempt as sabotage of the new state: anyone who withdrew would be held to have left the profession as such, and would consequently be regarded as unfit to farm German soil in future.[11]

The RNS had four departments dealing respectively with the farmer and his family as people, with farm management, with the co-operatives, and with the processing and retailing business areas. These four departments understood roughly the former administrative duties of the unions (no. I), the Agricultural Chambers (no. II), the co-operatives (no. III), and business associations (no. IV), as obviously the processing and trading firms had previously had their own professional organisations. Geographically, the lowest level was the *Ort* headed by an *Ortsbauernführer* (OBF) of whom there were just over 50,000, not provided normally with an office staff. The next stage up was the *Kreis*, each of the 514 being a *Kreisbauernführer* (KBF); at this level the four departments system began. The penultimate stage was a *Landesbauernschaft,* of which there were 19 initially: the leader was a *Landesbauernführer* (LBF). Finally at the top of the hierarchy was the national *Reichsbauernführer* (RBF) in Berlin with his deputy; these two posts were held by Darré and Meinberg, respectively. In addition there was a

Staff Office, headed by Dr Reischle, to plan and co-ordinate all activities. In some areas it was found necessary to insert another level between *Landesbauernscahft* and *Kreisbauernschaft*, known as *Bezirk.* Westphalia, for example, had six to cover its 37 *Kreisbauernschaften.* All leaders were appointed on the Führer principle, that is, by their immediate superiors, with no question of democratic election from below.

In theory this was an autonomous body rather than an appendage of the state. The NSDAP emphasised that it had erected a halfway house between state socialism and the untrammelled freedom of the individual allegedly entailed in liberal societies. However, in June 1933 Darré himself wrote an article in the *NS Landpost* stating that 'in the coming development it is the duty of the leader of the farmers to lay the politico-economic needs of the profession before the state leadership . . . the politico-economic decisions taken are within the province of the state leader, to whom the profession's chief can only ever be an adviser'. This asserts the clear primacy of political, that is state, considerations over those of professional representation in Darré's mind; it must be remembered that in any case he was eventually both RBF and Minister. Under these circumstances the claim that the RNS was autonomous can scarcely be taken as valid. Rather it was based on the NSDAP principle of an organic state, where each profession served the nation.

There were in reality three choices confronting German agricultural organisation at the time; it could have stayed with the old, pluralistic forms of the Weimar Republic, anathema to so monistic a party as the NSDAP. Two possibilities were left, a democratically-run unitarian body or an authoritarian one. In other words a real alternative was possible which would have combined both unity and democracy, as von Rohr suggested at the time. He too was dedicated to the idea of 'Blood and Soil' and the preservation of the peasantry, but he proposed a rather different way of achieving it. This was based on a vertical structure from district upwards with the highest level as a co-ordinating organisation only, rather than one which imposed its own policy as the RNS did. In combination with a democratically-elected leadership, which von Rohr advocated, this would have constituted a real alternative. It was no doubt the lack of centralised control in this idea which led Darré to reject it in favour of the RNS as eventually organised.[12]

However, as at this time there was a general move in western Europe towards a greater degree of state intervention in agrarian matters – for example, the Milk Marketing Boards in Britain – and it might well be held that what the RNS really represented was part of a general trend. It is true of course, that the NSDAP went further along this particular path than Britain without ever attaining the total state control, at least before 1939, exercised in the Soviet Union, so that to a certain extent the claim to have reached a compromise has some genuine grounding. Moreover, the move to fixed prices was a general feature before the war, as witness the state guarantees for wheat which the Popular Front introduced into France. In the narrower political sphere, the RNS represented a victory for Darré and his associates over the DNVP-orientated von Rohr, and the state control ideas in some sections of the NSDAP itself. Here again the claim to have found a middle solution has some genuine support.

Reorganised marketing was complemented by the Law of Hereditary Entailment (*Erbhofgesetz*). This applied to holdings only within certain size limits, and affected 55 percent of all agricultural land in the country.[13] The preamble stated that the object was to preserve the peasantry under the safeguard of old German inheritance customs. The chief weapons used were restrictions upon testamentary freedom and the total exclusion of any form of foreclosure by creditors. No land could be sold, and this fact, in concert with the inheritance regulations, meant that the soil would remain in the same family line in perpetuity. In this way that section of the programme of 6 March 1930 was fulfilled, which stated 'German soil will not be allowed to be the object of financial speculation' (although nearly half had not been encompassed in the new law). The peasants' family had now received both absolute security of tenure and guaranteed prices for their products: a privileged class, sheltered from all the worst effects of free market forces, had been created in Germany. (The fixed price measure applied of course to all food-producers equally, not merely to the peasants). Details of the new legislation are given in a later chapter, but some facts about the background of the entailment law in general here will help to place it in the context of German agrarian development.

There had been a groundswell of agitation in many circles in the twenties for some reform of the laws of farm property inheritance. The question was complicated by differing local customs; broadly speaking, there were two main systems of land inheritance. In the first (*Aner-*

benrecht) the farm went over intact to the sole heir: the other (*Realteilung*) consisted rather of partible inheritance. This was widely practised in the Rhineland and much of the south-west. Its obvious disadvantage was the fragmentation of the farm which it involved, as all members of the family got a share.

What this could mean was shown in one case in Württemberg in 1925 when the farm went to eight surviving children. Each heir received one strip, four more were to be cultivated in common, from the next four the eldest child got two and the others shared the other pair equally. Finally, the last four allotments were divided in an even more complicated fashion.[14] The real trouble lay in the non-viability of such parcels of land, especially because of the cumulative effect. A 1938 report on two parishes in the same region found the number of holdings under 5 Ha in size in one to be 296 out of 309, and in the other 183 out of 186.[15] Not all of the same peasant's strips were necessarily adjacent, so that he could face the problem of having to drive his cows three miles to pasture.

Unfortunately the German Civil Code of 1900 had sanctioned *Realteilung* in effect by not distinguishing between movable and non-movable farm property. What reformers wanted was a system whereby the farmer's children could inherit money equally, but which would retain the holding as an entity. A farm union in Bavaria requested the government for something similar in 1922 and again five years later.[16] In 1925 the leading farmers associations actually framed a national bill to this effect in conjunction with the Agrarian Council, called the *Reichsanerbengesetz* and sent it to the Cabinet for approval. The government took the matter up with the individual regions, but it transpired that at least in some cases they would prefer to introduce their own legislation.[17]

Several in fact did so. The usual basis was to draw up a roll containing the names of all those local peasants volunteering to nominate one heir only for their land. For example, Württemberg brought in such a law in 1929, and which incidentally was warmly welcomed by the leader of the NSDAP group in the Landtag. As well as voluntary enrolment the bill allowed farmers to retract at any time when they wished. The eldest son would take precedence as sole heir: for the other children there would be monetary compensation spread over ten years. For disputes over the true value of the property an arbitration procedure was laid down. It is important to note that sale of the holding was not precluded.

The entire object of the legislation was to reinforce *Anerbenrecht* in the area but in a purely voluntary way, as the Civil Code of 1900 was held to be injurious to the maintenance of a 'sturdy and capable peasantry' holding its land in perpetuity within each family line. It was pointed out that such a measure was already in existence in Prussia. Other regions already had similar legislation in the late 1920s or were in the process of preparing it, for example, Saxony and Lippe.[18]

NSDAP attacks on existing inheritance laws have to be seen against this background; for example, point 19 of the original party programme calling for 'Germanic law' to replace the 'materialistic' Roman law of the 1900 Civil Code. Reform was strongly advocated in Willikens' article in 1930 in the *NS Jahrbuch,* as well as in the March 1930 programme for agriculture. The NSDAP, however, differed from the regional policies already described by demanding both national uniformity, not individual state legislation, and a compulsory system of *Anerbenrecht* as opposed to the purely voluntary basis of regional laws. Its policy was thus only to some extent in German tradition; apart from Hanover prior to 1866 no single region, let alone the whole country, had ever before had an enforced system of closed inheritance.[19]

Consequently, when in 1933 the NSDAP Bill was presented it is hardly surprising that the Prussian Minister of Finance should have called it 'a step of immense fundamental significance'. It envisaged obligatory restrictions on testamentary freedom for the peasant, and also departed from previous ideas by making it impossible to foreclose on an entailed farm (*Erbhof*). Nor could the owner sell it outside the family line or offer it as collateral against a loan: he could not even dispose of any part of it without court permission. As one critic at the Cabinet meeting said, a new kind of feudal tenure had been created, with the whole nation as the lord, and the peasant as the tenant. Criticism of the concept either in principle or in detail came from several non-party ministers, such as Schwerin von Krosigk, the Finance Minister, or the Prussian Finance Minister, Dr Popitz. Hitler himself rejoined that the law was absolutely necessary in order to protect the peasant against land speculation, the result of which could be loss of the holding; on a national scale this would mean ruin for the country. The peasantry were the foundation of the people and without it the German population would be reduced to thirty to thirty-five million only in little more than a generation.[20] After this intervention the measure was adopted. The point about population statistics will be taken up in a later chapter but it is informative to see this advanced as a criterion by Hitler of the need to maintain the peasantry.

Together with the *Reichsnährstand,* the new law reformed the two pillars on which German agriculture as a whole was to be rebuilt, in particular the peasantry, as the party had promised that it would be. In his presidential campaign address to the farmers in April 1932 Hitler had reiterated Germany's need for self-sufficiency and that the NSDAP did not see its future solely in terms of international trade. On the contrary, 'The true welfare of our people does not seem to me to be expressed in terms of import/export statistics but rather through the number of viable peasant holdings', also seen as the best safeguards against social evils and racial degeneracy. When one farmer in Schleswig-Holstein was asked why he had voted for the NSDAP prior to 1933 he said it was because in the Third Reich farmers would be strong enough to get what they wanted.[21] Even if that still did not happen in autumn 1933 it does not alter the fact that substantial gains had been made by the agrarian community.

The net effect of the economic changes was striking, as shown by the statistics in Table 1.[22]

Table 1

Year	Farm sales	Expenses	Surplus
	(thousand million RM)		
1928-29	10.2	8.6	1.6
1930-31	8.6	7.4	1.2
1932-33	6.4	5.9	0.5
1933-34	7.4	5.9	1.5
1934-35	8.3	6.0	2.3

The advent of the new regime had given the agrarian sector a substantial increase in income represented by the surplus of sales over expenses. This was in stark contrast to the continual decline which it had suffered in the last years of the Republic, under the lash of the world economic crisis. What is especially interesting is the way in which this was achieved, not merely through higher prices for produce, but also by holding expenses steady. Debt-relief measures and tax cuts were mainly responsible. In 1932-33 interest payments and taxes represented 22.1 percent of total farm sales, in 1933-34 the corresponding proportion was 18.8 percent and in the following financial year only 13 percent. These figures reflect both rising incomes and cuts in taxes and interest rates: in 1934-35 the total of both these latter items was less than two years previously in absolute terms.[23] In sum, at least some expenses fell whilst incomes rose. Moreover, as taxes and interest-rates had been the main burden of financial complaint amongst the peasantry, those reductions were no doubt of particular psychological significance.

Moreover, agriculture was now doing better financially than other sectors of the economy. Eventually this situation was to change, but in the financial years 1933-34 and 1934-35 the annual percentage increase in agrarian revenue was 17.1 and 16.9, respectively, for allother sectors combined the figures were 6.8 and 12.2 only.[24]

Statistics of this kind, in conjunction with judicious flattery, and legal reform, enabled the NSDAP to consolidate its hold on the whole agrarian scene. To the landed community the Third Reich was fulfilling its promises, more to the peasantry than to other sections, admittedly. It was not, of course, the case that the party was anti-industrial, but simply that the abolition of reparations gave it the chance to initiate measures that would have been difficult before that. Germany still had to export to some extent as Hitler pointed out in the Reichstag in March 1933. But there was new room for manoeuvre after June 1932 that could be utilised; the largely DNVP-inspired programme followed by Hugenberg and von Rohr between January and June of the following year had already shown this.

Party speakers rammed home the propaganda advantage of this gradual change by habitual reference to a 'peasant policy', as Meinberg described it. In a speech in September 1933 he referred to a conversation he had had with Darré and Hitler a few days previously, at which the Führer was alleged to have said, 'I shall not yield an inch in any question affecting the peasants'. Meinberg went on to stress what this meant: price increases in other branches of the economy would not be tolerated if they adversely affected the balance between agrarian and other wholesale prices for the peasants. Any attempt to do that and 'the band of criminals' involved would be put into a concentration camp, a statement which seemed plain enough.[25] The Agrarian Office had already adopted a similar line in its approach. In February 1933

Darré called on the party's press to make widespread use of the phrase 'Peasant Chancellor' when referring to the Führer, as he called him at the Harvest Festival ceremony in October of the same year, just after the *Erbhofgesetz* was announced. Hitler himself never lost the slightest chance of assuring the peasants of his goodwill towards their profession; on 1 October 1933, for example, he received a delegation of 100 at the Chancellery and emphasised the firm bonds between his government and agriculture.[26]

In a wider context the differences between the measures of the DNVP policy up to June 1933 and Darré's own legislation afterwards are as revealing as the similarities so that continuity should not be exaggerated. Outwardly there was sometimes little to distinguish them; von Rohr believed in 'blood and soil' and the intrinsic value of a sound peasantry just as much as did Darré. Equally, many agrarian associations had been calling for inheritance law reforms. None the less, the new policy was distinctive both in terms of means and of ends.

Both in the *Reichsnährstand* and the new entailment laws there was a clear authoritarian element lacking elsewhere. The obligatory membership of the RNS, based on the Führerprinzip and an hierarchical structure, was paralleled by the testamentary restrictions of the Erbhofgesetz. On the other hand, the NSDAP could clearly be given credit for the internal coherence of its programme. Here there is a very clear and striking contrast with the rather patchwork medley of policies of previous governments. Economically the Weimar cabinets had really rather drifted with the tide, or in some cases, such as that of the Maize Monopoly, been pushed into some kind of expedient or other to meet the exigencies of the moment. Perhaps one of the most revealing facts about the NSDAP's policy towards agriculture was that it could write the word 'programme' in the singular at all.

The actual values inherent in the latter reveal starkly the utterly conservative, anti-liberal, NSDAP attitude towards society in general. This view was by no means confined to the Agrarian Office but represented rather the central core of the National Socialist outlook, as witness the 1933 speech by Goebbels in which he pronounced the year 1789 as now having been removed from history. Much ink has been spilt in attempts to determine whether the NSDAP represented the last stage of monopoly capitalism or whether it was itself anti-capitalist. This is certainly an over-simplification, judging from its agrarian policy. In reality, it simply depends on how capitalist is defined. The NSDAP was certainly not opposed to private property or to private initiative in the realm of business. Hitler himself pointed out in the Reichstag on 23 March 1933 that Germany would not become the field for economic experiment, nor would any bureaucracy be constructed to deal with the present crisis. The conservation of the peasantry obviously implied the continuance of private property in agriculture likewise. The NSDAP here was as tactically pragmatic as always.

But when the ultimate aims of the new agrarian legislation are considered, they can hardly be said to have been conducive to any further developments in capitalism on the land. First, the peasantry was one link in the national chain only. Very clear duties were laid upon it in return for its protection, as many articles published in the Third Reich underlined. The entailed farm was run very much in the service of the people as a whole, rather than merely as an object of private profit, although it was accepted that the peasant had to think of that too. But primarily he had responsibilities as the source of the race, as the backbone of national defence and as the man who fed the nation. This very tight link between duties and privileges is more reminiscent of the medieval era than of any liberal bourgeois society, as indeed is the whole concept of a people as a kind of living organism. In this respect the NSDAP was profoundly anti-capitalist, because of opposition to liberalism in the economic sense, which put its emphasis on the individual, whereas Marxism put emphasis on the class; the core of National Socialism was the race. For the NSDAP, as Hitler said at a peasant demonstration in September 1933, 'the individual is transitory but the race remains'. It is really rather hard to see how in the long run such concepts could be anything but inimical to liberal capitalism.

Equally, the *Erbhofgesetz* quite clearly aimed at the maintenance of the existing numbers of peasant holdings; the elimination of free market forces was bound to entail this. To lift the peasantry out of the field of their operations was ineluctably to fix it like a fly in amber at the current stage of its development. For ideological reasons there simply could not be fewer peasants in Germany in the future, however more rational and efficient that might make production. As Darré put it, there was no room for 'grain factories' in Germany.[27] Neither collectivisation nor a monopoly of large-scale private estates could be accepted. This was the inevitable consequence of putting racial considerations before the purely economic. Darré never missed a chance of hammering home this point to his subordinates. For instance, he told the LGFs in February 1935 that whereas in the Soviet Union and in Fascist Italy agrarian policy was based mainly on the desire to maximise production through rationalisation, in the Third Reich the overriding issues were blood and race.

Given these priorities, the question of whether the party was anti-capitalist or not simply does not arise in any meaningful sense. It permitted private property and private initiative where they did not threaten the race; it could not consider as acceptable any aspect of liberalism, in economics or otherwise, which did. Ultimately, it is rather hard to see how its legislation for agriculture was compatible with the operations of free enterprise in the long run. As one historian has said, 'Feudalism had as much to do with Fascism as did capitalism'.[28] Both the corporative state and National Socialism were reactions against the modern world of liberal bourgeois democracy. Nowhere is this clearer than in the agrarian policy of the NSDAP. Hence when Herbert Backe said at an agrarian assembly in 1934 that Germany had gone from a free economy to an organic one based on obligations (*gebundene Wirtschaft*) he summed up the NSDAP's agricultural policy and its whole outlook in one phrase.

A brief description of the *Reichsstellen* completes the picture of marketing reorganisation. Unlike the RNS, the latter were set up originally to supervise imports of foodstuffs. As early as December 1932 there was an office for fats, which was quickly followed by the erection of similar bodies for grain and fodder and dairy products, meat and livestock. Any one bringing such commodities on to the inland market from abroad had to offer them for sale to the appropriate *Reichsstelle,* which had the right of refusal. In that case, the goods could not be offered for sale in Germany. These regulations complemented the RNS in effect, and permitted the authorities to maintain a close supervision of all transactions on the agrarian market. By the late 1930s the *Reichsstellen,* which actually dealt in commodities directly, were coming to intervene more and more upon the domestic market. They were also furnished with their own departments for issuing permits for the use of currency when food purchases were made abroad.[29] The whole question of foreign trade in the Third Reich, and its relation to agrarian policy, will be dealt with later, but this brief summary of the import boards is given here to show how the RNS control of the inland market was complemented by that of foreign purchases.

6

THE REICHSNÄHRSTAND IN DETAIL

An Analysis of its Personnel and Their Duties: The Reorganisation of Agrarian Marketing and Prices and Incomes Policy

As the head of the RNS was a National Socialist this meant that the entire hierarchy was drawn overwhelmingly from the party ranks, because of the Führerprinzip system, whereby each leader at whatever level appointed his subordinates. Most of the leaders in the RNS had formerly been in the Agrarian Office and simply went over to the new body en bloc. The existing LGFs now became LBFs, i.e. the heads of the nineteen different regional branches of the new organisation as an examination of the LBFs in 1934 makes quite clear. They were all professional farmers, ranging from estate-owners like Freiherr von Reibnitz in Silesia to peasants such as Hubner, the LBF in Baden. This applied equally at the lower levels of the hierarchy. Despite the use of the word *Bauer* in their title these men did not actually have to be proprietors of an *Erbhof,* although they did have to be engaged in agriculture for a living as landowners of some kind. In sum, the *Bauernführer* who carried out RNS policy were professional farmers and almost always party members.

Their white-collar staff tended to be highly professional, since Department I of the RNS consisted mainly of the old unions' personnel, in particular of the *Landbund,* which had been used as the basis for the new structure created in April 1933. When it was incorporated into the RNS the existing professional agrarian officials came over automatically. The same applied to Department II of the RNS which comprised the staff of the former Agricultural Chambers. Thus the new organisation tended to consist initially of NSDAP *Bauernführer* as the leaders, while the subordinate personnel were frequently not party members, but rather professional agrarian administrators. One contemporary leader has estimated that up to one-quarter of the officials in the Staff Office of the RNS were not in the NSDAP.[1] This claim has independent evidence for its support: as late as May 1937 the *Landesbauernschaft* in Westphalia had fifteen officials who had formerly been in the Centre Party, nine of whom were in agricultural training schools.[2] Despite the initial prevalence of non-party members there does appear to have been pressure exerted later to produce a wholly NSDAP structure, even for office staff, as is clearly shown in a letter from the Rhineland *Landesbauernschaft* to its personnel virtually ordering it to enrol in the party by a certain date.[3]

It seems very likely that the organisation became in general increasingly dominated by party members in the course of time. Of course, the very fact that staff were pressurised into joining the NSDAP means that they were not necessarily convinced National Socialists. The main impression given is an air of professionalism: it was certainly not in general staffed by National Socialists in virtue of their political allegiance. The smooth running of the new organisation in Schleswig-Holstein has been attributed to its solid professional core.[4]

Of course, almost inevitably some unsuitable people did get nominated to positions of responsibility. For example, a KBF in the Osnabrück area was said to have been neither a farmer nor a local man and of inferior ability; from Münster came a report that a local *Bauernführer* was an old party member who had married a farmer's daughter but otherwise had no discernible connection with the land.[5] There is no doubt, however, that many senior officials felt the need for a professionally competent leadership of agriculture at local level. At the National Peasant Assembly in 1935 Dr Krohn, a section-leader of Department II at headquarters, argued that the peasant only allows himself to be convinced by other peasants, so that if progress were to be made technically the OBFs would have to be highly competent at farming. This was certainly not always the case. As late as 1940 one OBF in Marienau (Lower Saxony) was said to be technically unreliable and had been the object of complaints from the local peasants.[6] Reports of this nature were, however, relatively rare and seem to have designated the exceptions rather than the average: as the RNS developed it tended to grow more professional in its appointments and less politically biased, since it was found by experience that the peasants simply would not co-operate with *Bauernführer* whose only qualification was party membership.[7]

Quite apart from professional knowledge of agrarian matters, all *Bauernführer* were required in some degree to be administrators: since they were farmers by profession they had no special qualifications in this respect, which was initially a weak point in the RNS as a whole. An examination of the correspondence of OBFs shows that many were only semi-literate; in Schleswig-Holstein they were described as being of variable quality.[8] One example of the OBF's lack of education in general was over the speech at harvest time, which the OBF was called on to make when the first wagon arrived from the fields in each village, at least in Württemberg. Most were quite unable to compose this and bombarded the *Landesbauernschaft* for instructions.[9] Their inability to manage this on their own behalf suggests that, however skilful as a farmer, the average OBF was not an educated man.

This obviously was of importance because of the vast amount of administration which the local leaders were called upon to undertake, and which understandably involved a great deal of paperwork. For the OBF the burden of administration was so heavy that Darré ruled that no one should remain in this office above the age of fifty-five, when he received an honorific 'Old *Bauernführer*' upon retirement. When their list of duties is surveyed, it becomes clear why this regulation was necessary. Just one set of orders detailed the OBFs to try and sell copies of an RNS calendar, find out how many *Erbhöfe* in the district needed debt-relief, and complete a questionnaire regarding the supply of timber from private forests.[10] At a 1938 referendum they were expected to participate in the campaign for the party and assist with the machinery of vote-registration; when it was decided to install more bath facilities on the land, it was the OBF who was charged with the execution of the plan in his village.[11] The list of duties was almost endless; additional tasks included checking grain deliveries at every farm, seeing that eggs were delivered to official collection points only, administering state aid for technical improvements such as silos, reporting on land helpers sent from the cities, ensuring that farmers' wives joined the party organisation *NS Frauenschaft*, dealing with land-workers' accommodation, checking cattle diseases, publicising new RNS reading material, and reporting any peasants meetings. Even that was not the full monthly list for a man who had his own farm to run and was not even paid for the work, in addition to having no office staff. Moreover, a conscientious OBF regularly visited all holdings to urge on better management.

Under this kind of pressure it would have been surprising if all duties had been carried out as efficiently as possible. Apparently, however, supervision of the peasantry was sometimes laxer than the RNS would have liked. In Leezen (Schleswig-Holstein) for example, a peasant was deprived of the management of his holding and his wife installed as trustee in March 1937; over two years later the KBF found out quite accidentally that she had died months before and that her husband was back in control.[12] That he could have been managing the property in defiance of a court order without the OBF apparently knowing about it is informative. Sometimes they were in doubt about the marketing regulations of the RNS, judging from a case in Lower Saxony in 1937 when the local *Bauernführer* had apparently failed to check transgressions of the cattle price.[13] Of the paperwork required in general one peasant leader reported that there was so much to do that it made the authorities look ridiculous.[14]

The *Kreisbauernführer* were also made to earn their money at the next link in the chain. One in the Rhineland worked six days a week for the whole of one month taken at random from his diary and on one of the Sundays as well; in addition he often had evening meetings.[15] The trouble was for the KBFs that quite apart from all the multifarious official duties the peasants inevitably came to regard them as a kind of general factotum: no doubt the fact that they were usually party members strengthened this tendency. The local farmers assumed naturally that they exercised some influence and consequently were the people to consult in an emergency. So when one peasant's sister was left some money in her father's will and the heir to the farm did not pay up it was to the local KBF that she went for redress. Equally he might be called in for local debt-collection as when a cattle-slaughterer owed money to a peasant.[16]

When all the innumerable administrative tasks involved in agriculture, food-processing, and distribution fall upon one body, ineluctably it will become cumbersome. The very size and apparent unwieldiness will give observers an impression of inefficiency and red tape. However, it has to be remembered that the bureaucracy of smaller units is less easily noticeable. Clearly this has to be borne in mind when assessing the size and cost of the RNS as it comprehended the former farm unions, the Chambers and the wholesale and retail food trades. Consequently by June 1938 there were some 73,000 employees in the RNS, although of these only 16,000 were paid, because the 55,000 OBFs were honorary.[17]

Criticisms about the expense of the new corporation were soon raised, especially by local government officials. In fact, the point was made in so many quarters in Prussia that Goering (as Minister President) had them collated and sent to Darré for his information and comments.[18] The sheer size of the RNS made it costly and as it depended for its own internal expenses on the contributions levied on individual firms or food-producers the sums demanded tended to be high: by 1934 the administrative costs amounted to RM 69 million. The Ministry of Finance was quick to suggest that a poorly-organised, over-elaborate structure was generating unnecessary expenses:[19] it was clearly felt that the whole corporation was demanding too much from its members. A food-processing firm in Königsberg, for instance, claimed that it had been asked to pay RM 1,200 as an annual subscription, whereas prior to the RNS it had paid less than a third of that amount to professional associations. Similar considerations applied to peasants. From south Germany it was alleged that three farmers were contributing between them nearly four times their former amount. It was even said in 1935 that subscriptions were so high that they actually were harming the export drive.[20] This sounds like an exaggeration but clearly from its size and its outlay the RNS gave an impression of bureaucratic unwieldiness.

The basis of the actual marketing reorganisation was the individual commodity, for each of which a separate structure was built up. These were administered at first by Departments III and IV of the RNS, which dealt with agricultural dealers, the processing firms, distributors in general, and the co-operatives. All of these interested parties were represented in an association for each leading agricultural product, commencing at regional level, where the new body was called a *Verband* (plural *Verbände*). At the top was a national Federation, of which ten were eventually organised, in respect of grain, livestock, milk, fats, potato flakes (for fodder), eggs, spirits, sugar, fruit and vegetables, and fish. For a number of other products commissioners were named pending the formation of a Federation. The RNS could depute all its powers to the *Verbände* if it wished, but these bodies were none the less supervisory only: no *Verband* or Federation was a trading concern in the sense of buying and selling goods. These latter functions continued to be exercised by co-operatives or by private dealers. In 1938 the grain harvest was bought by private dealers, millers and co-operatives in a ratio of 49 : 16 : 35 so that dealing was left to private initiative.[21] The proportion handled by each type of buyer had scarcely changed since 1935. What the *Verbände* did was to regulate prices and marketing conditions, license to trade, etc. – in other words to administer the market without entering into its actual transactions. In cases of dispute over its decrees, a Court of Arbitration for each commodity was responsible, dealing not only with cases of transgressions against the rules but also in those where, for example, a peasant came into conflict with a co-operative or miller. If a business was pronounced no longer viable and closed by the *Verband*, it was to this court that the owner had the right to appeal.

In order to make the system clearer the grain trade will be taken as an example: this was brought into a unified association on 17 July 1934.[22] The Federation headquarters in Berlin had nineteen regional groups under its wing. The Federation chairman was nominated by the Minister for a two year period of office, and provided with an administrative council of at least eleven members from the regional bodies themselves named by the Minister in agreement with the RBF. The lowest rung of the ladder was the district *Verband*, whose chairman was nominated by the LBF in agreement with Federation headquarters; each *Verband* council had at least nine members, similarly chosen. Peasants, millers and bakers each had two of these representatives, whilst one member each from the flour and grain dealers and the local co-operative made up the nine. Each district *Verband* had as a supernumerary advisory council an assembly of thirty members, of whom ten were peasants or farmers.

The duties of the chairman were as described already for *Verbände* in general. As an example of their powers over delivery, the regulation of July 1936 decreed that all farmers must deliver at the mills by 15 October an amount of grain equal to 30 percent of the previous year's crop, with a similar proportion to be sent in afterwards, but before 31 December: the decree applied to all producers with more than five Ha of land. Behind the staggered delivery system was the desire to bring the crop on to the market at a steady rate, and so avoid an early glut and then a subsequent shortage, as might occur under a free delivery arrangement. Prices were varied from month to month, as an added inducement to regular deliveries, and also from region to region, according to the distance from the main market being supplied; in May 1936 the price for rye ranged from RM 163 to RM 183 per ton according to this system. Price and delivery rules thus complemented one another in assuring a steady flow, and thereby security, to the consumer. Equally, the producer now had a guaranteed price for his grain. The new arrangements also permitted compensatory payments within the commodities handled; in 1935 the result of a poor rye harvest was balanced by a levy of RM 6 per ton on wheat.[23]

As might be imagined, the regulations involved a fair degree of bureaucracy, since clearly any such close supervision would require a great many supervisors. The Federation originally set up for the millers as an independent branch of the grain trade had an office staff of nearly 500 at its headquarters in June 1934, which two years later had virtually doubled in extent. An accountant's report in the following year described the entire apparatus as over-manned and spoke of delays in procedure, too much formal organisation and uneconomical management.[24] The intricate nature of the regulations makes it easy to understand why there were such problems of control and why the laws were not always understood on the land. The original decree establishing the new order ran to no fewer than eighty pages in the official gazette, a hard nut to chew for relatively inexperienced administrators.

A great many claims were made by the new corporation as to how it had rationalised the whole distribution complex, in contrast to previous practice. As a quoted example of free market arbitrariness, milk from the Allgäu in southern Germany had been sent to Berlin, resulting in unnecessary freight charges: the RNS sorted this out by establishing fifteen distribution zones to facilitate its transit from producer to consumer.[25] As a result of the (allegedly) better system there was a vast increase in the quantity of milk delivered to dairies: in Württemberg this was claimed to have nearly doubled between 1932 and 1935.[26] Actual production was also said to be stimulated by differential price controls; in the Eifel (Rhineland) area the farmers had almost given up milk production prior to 1933 owing to their remoteness from the market: especially favourable prices were offered to them by the RNS as an inducement and a substantial increase in output was achieved.[27]

Parallel with such developments ran the efforts at rationalisation which entailed pruning out the number of middlemen in each commodity trade, by making use of the Federation/*Verband* powers to license to trade or to close down a business where that seemed to be in the public interest. There were 10,000 dairies in Germany in 1933 but by 1939 only just over 6,000 remained, the smaller concerns having been eliminated in the interests of efficiency.[28] Failure to achieve a certain given trading quota meant closure, or amalgamation with other business firms in the same line, only three months' notice being given.

It is interesting to compare these points with comments made about the RNS by food-producers, both in respect of the arrangements regarding distribution and of prices. The first evidence in this respect was provided by the *Regierungspräsidenten* reports of July 1934, thirty-one in number, sent to Goering in his capacity as Minister of the Interior in Prussia (see note 18). The first sore point regarding the new regulations concerned the general handling of produce, in particular the fact that peasants could no longer in all cases sell direct to the public but had to deliver their milk to a dairy, which also entailed being unable to make their own butter. This point links with claims about how the RNS had increased milk deliveries; this was scarcely surprising if peasants had to send it to dairies anyway. In other words, official propaganda had simply concealed that increased deliveries to dairies did not automatically mean that production had been augmented. So strong was feeling on this particular point that the peasants in the parish of Egge (Lower Saxony) initially refused to supply the nominated *Verband* with milk at all, which led to the KBF talking about 'sabotage'. The root of the matter here was that the locals possessed all the necessary apparatus to make their own butter, which they were now forbidden to do in effect. Their expensive equipment was now useless to them, which makes their resentment understandable.[29]

RNS intervention in the sales process also caused friction in the case of other commodities. From Aachen the Gestapo reported that producers were annoyed at having to use the *Verband's* services in order to dispose of their potatoes. The whole question throws an interesting light on the new regulations and their effect; if peasants had been accustomed to selling direct it was irksome to have to go through an intermediary, and simply interfered with the free flow of goods. This interruption of normal trade procedures was almost certainly at the root of continuing reports of maldistribution of foodstuffs throughout the Third Reich, even in peacetime. Both from Wiesbaden and from Cologne the *Regierungspräsidenten* announced in July 1934 that eggs were no longer fresh when sold, due to the complicated marketing regulations. Egg deliveries seem always to have defied the RNS's edicts, as in 1937 there was talk of a black market in Bavaria.[30] That this should have been the case under otherwise normal circumstances is an interesting commentary on the RNS.

The claims for a more rationalised distribution appear exaggerated for other commodities. In May 1937 a mill near Jülich (Rhineland) had to be closed for six weeks due to a grain shortage in the area. In 1938 the OBF in Ruhpolding (Bavaria) claimed that no cattle-dealer came to the district for two to four months on end, which compelled farmers to feed cattle in winter which they had wanted to sell in the autumn. There were too many similar instances for such events to have been mere chance. In the Aachen district both fats and eggs were lacking in July 1936 and in the following year fodder shortages made it hard to find pork in the shops in the area.[31] An overall impression of maldistribution is ultimately difficult to avoid.

An early attempt to improve the marketing system was made by Darré in November 1934 when he called a conference under Backe's chairmanship for RNS officials and representatives of local government.[32] Dr Reischle admitted that a certain dualism in administration existed between Departments III and IV which would now be overcome by amalgamation. To improve marketing still further LBFs would in future be given unified control over its regulations. That these changes were not sufficient to produce a completely satisfactory distribution system can be gauged not only from examples after 1934 already quoted here, but from Darré's own judgement of the RNS in this field in 1936. In a letter to Goering he twice mentioned that the distribution of produce was the weak point of the organisation:[33] so impressed was Darré with this that he apparently feared that the old idea in the party of handing marketing over to the state might be revived. After such an opinion from its chief any further summary of the RNS in this respect would surely be superfluous.

The second main issue of importance in the new system was prices. Here there are two points to consider: the price rendered to the producer and its relation to the selling-price, usually called the price-span in the Third Reich. Farmers felt the first to be too low, according to the *Regierungspräsident* for Königsberg in July 1934; from Aachen the Gestapo gave it as a current complaint, especially for butter, cheese and eggs.[34] The peasants seemed to feel that one reason why their financial reward was so limited was that there was simply too great a difference between what they were getting and what the public paid. This problem seems to have persisted throughout the Third Reich, as in 1938 there were complaints from Bavaria on the same score. The OBF at Eggenfelden estimated that on a sack of wheatmeal a baker could make over 60 percent profit, which amounted to a larger sum than the peasant got for producing it.[35] In the Rhineland leaflets were handed round during the carnival in March 1938 containing malicious attacks on the LBF in Bonn and the RNS in general 'because the producers' price is too low in relation to the selling-price'.[36]

Indeed when Hitler himself read the *Regierungspräsidenten* reports from Prussia of the previous July, he demanded an explanation from Darré pointing out that he had been made aware on journeys throughout the country that the public were complaining about what they had to pay for bread, milk and butter. The Führer clearly found this hard to reconcile with accounts of farm complaints about low prices. Darré could only reply that reports of dearer food to the consumer had been exaggerated.[37] Even if that had been the case it appears evident that the RNS had failed to solve the problem of the gap in prices between shop and farm and from the 1938 reports cited here, it seems that this issue was never resolved over the short time-span under review. A similar situation had, of course, existed prior to 1933, but surely at least one goal of the RNS regulations had been to change that.

The real issue to be resolved here is where exactly to apportion the blame if farm prices were too low. After all, producers' complaints have to be seen against the background of the general economic situation of the time. Hitler had already made good his promise to raise prices to the farmers at the consumers' expense. Figures have already been quoted here to illustrate how well the peasants actually fared prior to 1935 at least. Under these circumstances, with mass unemployment, was it a justifiable attitude to ask for even more money? The agrarian community was back to the same stance that it had adopted before 1933, namely of making demands the burden of which would fall on others.

Ultimately, the answer lies in an analysis of the RNS as such, and its whole marketing structure and practices. If the new corporation, as an intermediary, was absorbing too much of the price-span by way of expenses, then clearly without it the producer could have received a higher price without hitting at the consumer. This was the crux of the matter. The *Verband* took a commission from the peasants which, in the case of milk, was 2 Pfennigs per litre. This was clearly an imposition for those producers who had hitherto sold direct to the customer. In one district at least this could equal as much as 20 percent of the farmer's profit.[38] This was unfortunate for those peasants whose leases were based on previous margins. But on the other hand many peasants, especially those living in remote areas, had not sold direct and therefore simply did not have the same cause for complaint.

This accounts for the wide discrepancies in the attitude adopted by the peasants to the regulations. The introduction of the new system in Schleswig-Holstein is said to have proceeded 'without friction'. Equally in the Hanover area the arrangements were apparently welcomed by the vast majority of producers.[39] On the other hand from Aachen it was bluntly announced after the beginning of the RNS that 'the mood on the land cannot be described as good'.[40] These conflicting accounts clearly indicate that local, or even personal, conditions varied too widely to permit any facile generalisations. From statistical evidence in general it can hardly be denied, however, that prices were more favourable to the producer after 1933 than they had been before.

On the other hand, the new marketing regulations seem to have been in some ways unnecessarily complicated and cumbersome in respect of actual distribution. Even this judgement has to be tempered by two considerations, however, since the Third Reich did not have a long life prior to 1939. Difficulties in the RNS might well have been ironed out in the course of a longer development period. Moreover, in judging scarcity in Hitler's Germany the point has always to be borne in mind that a rearmament programme in a country short of foreign currency is bound to produce strains and stresses, as will be seen in the later chapter dealing with autarky. There was not always enough to distribute.

One more issue needs analysis, namely, whether or not guaranteed prices actually aided German agriculture in the long run. This was thrown into sharp relief by the events of 1934-35, which provoked a lively battle on the whole matter and exposed Darré's corporation to some lively criticism; this crisis commenced almost as soon as the RNS was founded when adverse weather conditions affected the crops.

In contrast to the bumper crop of 1933 the grain harvests of the two following years were poor, especially in the case of the first. Unfortunately, it was hard to cover deficiencies by purchase abroad as German reserves were so low at this time; by December 1934 they stood at only RM 156.5 million.[41] This was clearly inadequate and as a result there were food shortages in the country which triggered off retail price increases, to which higher prices for farmers were already contributing. That the 1934 harvest would be bad was so obvious that even before it was gathered in Darré wrote to Hitler asking him for a 20 percent increase in the bread price, or some other form of compensation for grain producers. The Führer rejected the suggestion in a personal interview, stating that 'an increase in the price of bread must be avoided under all circumstances'.[42] Hitler almost certainly felt that as the peasants had already been dealt with generously they could not now demand more. In addition, he feared the consequences of dearer food on public opinion in Germany.

By late autumn, however, the situation was beginning to get out of hand as food shortages pushed up the retail food price-index which rose to 119.5 in November (1913 = 100). In January 1933 it had stood at 111.3 only.[43] This was too much and in Cabinet Hitler declared himself unable to tolerate such a state of affairs any longer. He accepted that he had set himself the goal of saving the peasants, but that he had also promised the public that wages and prices would not be allowed to get out of step. In such an event he would be accused of breaking his word, which could lead to a revolutionary situation. In this respect Hitler showed prescience. At a Trustee of Labour Conference in August 1935 exactly this point was made, the working-class expected the Führer's promise to be maintained.[44] It was Hitler's sensitivity to public opinion which led him now to demand Cabinet approval for the installation of a Price Commissioner as a kind of watchdog to oversee prices in those areas where supervision had formerly been carried out by the Ministries. In cases of doubt the Führer himself would decide from now on. At this point in the discussion Darré could only claim that he had a duty to make the German people independent in foodstuffs so must have a free hand with prices and production. He had no objection, however, to the supervision of middlemen by a Commissioner (whose appointment was to terminate in any case on 1 July 1935). The installation did not, of course, affect RNS legal powers of fixing prices for agrarian produce.[45]

The nominee was Dr Goerdeler who had performed a similar duty under the Republic. His appointment was welcomed by the press, one newspaper saying quite candidly that retail price control had been unsatisfactory before, as the RNS, local authorities, and the party had all had a hand in it. The article gave an impression of relief that at last one man alone had the task.[46] Geordeler at once issued instructions for meat prices to be held firm. Where they already seemed too high they could be lowered by agreement between himself and the RNS official responsible for meat and livestock marketing. By December he was recommending a cut in margarine prices to Hitler; both Schacht and the Finance Minister agreed, despite protests from Backe, who feared that lower prices would mean higher demand, unacceptable in view of current fats shortages.[47]

Other friction had already occurred between the RNS and the Commissioner. Goerdeler gave a press conference just after his appointment, and stated that supply and demand was responsible for price increases in recent months, plus aid to agriculture, which he accepted as unavoidable. He then cut rather close to the bone, as far as the whole RNS ethos of fixed prices was concerned. The concept was false, he said, because by eliminating competition, guaranteed prices merely helped the inefficient. The Commissioner then went on to say that production-costs in agriculture should be lowered by an examination of the distribution expenses. In effect this was open war against the RNS, which quickly defended itself in press articles a week later. No one was actually named, but it was stressed that advocates of a free market system should have the whole party against them and not merely the peasants. Swiss newspapers reported an open feud between the economic radicals in the NSDAP around Darré, and the exponents of old-fashioned economic liberalism, especially Goerdeler and Schacht.[48]

Darré knew the latter to be his real foe. At the 1934 Peasant Assembly he made a powerful speech aimed at Schacht, denouncing economic liberalism and defending the whole ethos of the RNS marketing system as such. In 1936 he described the Minister of Economics as a ringleader of the opposition to Hitler since 1932, who had been using the Price Commissioner as a weapon against the RNS whilst staying in the background himself. His aim was, allegedly, to shake public confidence in the RNS as a preliminary to destroying it, and eventually National Socialism in Germany altogether.[49] Whether this Machiavellian project was really in Schacht's mind is doubtful, but certainly there was no love lost between him and the Minister of Agriculture. The currency battle was partly responsible but Schacht's opposition to the RNS as such was real, because of his belief in the economics of free marketing. It was almost certainly Schacht that Darré had in mind when he told Hitler in 1936 that current difficulties over foreign trade and the whole economy were due not so much to the measures undertaken as to the personalities of some of those responsible for them. They had been chosen during the Weimar period and were dedicated to economic individualism.[50] His bitterness was attributable to the 1935 crisis, in part. Darré was bearing the brunt of the criticism for shortages, to which import limitations were also contributing.

All this can hardly have added to Hitler's appreciation of Darré and the RNS (since he was kept fully informed of events).[51] Goerdeler's attitude certainly did not help, because shortly after his retirement from the Price Commissioner's post in the autumn he gave the Führer a memorandum setting out his views on price control in principle. Like his press conference, this amounted to a sharp attack on the whole RNS fixed-price concept, and he reiterated his opinion that it merely sheltered the less efficient.

Exposed to criticism over a period of some six months, the RNS had to defend itself. In particular, Meinberg took up the battle in an article.[52] This was based on the proposition that guaranteed prices had originally worked to bring about recovery for the producer; but now, at a time of shortages, they principally helped the consumer. The deputy RBF argued that under supply and demand conditions when food was scarce prices would clearly rocket. Without the RNS the German public would currently be paying RM 500 million more for the same volume of food. The tenor of Meinberg's composition was equality of sacrifice: in 1933-34 the public had been asked to help the farmer, now the agrarian sector's income was in effect being curtailed to assist the consumer, and indeed the whole economy, which benefited from relatively stable prices.

These are cogent arguments when seen from the standpoint of the community at large. Clearly guaranteed prices were now preventing the farmers from cashing in on scarcity, as they might otherwise have done. On the whole the 'equality of sacrifice' line of thought seems to be virtually irrefutable, as undoubtedly in a supply and demand situation the whole industrial recovery of the country could have been jeopardised if inflation had developed. There were still two million unemployed at the time.

However, this line of argument by Meinberg in effect contradicted Darré's own request in 1934 for higher bread prices. On the whole, the former's reasoning seems more valid. The RNS leaders and planners had always claimed that their system had sheltered the peasantry from the dangers inherent in free marketing. And yet as soon as supplies became short Darré immediately asked Hitler for higher prices. This was to have the worst of both worlds with a vengeance. The RNS protected producers under a general guarantee, thus preserving the incompetent along with the vigorous; then at a time of scarcity there was in effect an immediate reversion to the laws of supply and demand anyway, with the usual implications in such a move of harm for the consumer.

The 1934-35 crisis seemed at first sight to justify Goerdeler's contention that to have followed the ordinary laws of free market operation from the first might have been better in the long run. The main objection to this from the NSDAP standpoint would have been socio-political as well as monetary. Darré simply was not prepared to let even inefficient peasants be eliminated by economic forces. Hence the attempt first to get more money but still stick to the guarantee system, and only when this failed, to produce slogans about equality of sacrifice. That is why the Meinberg line sounds suspiciously like making a virtue out of necessity; only after Hitler refused price increases did the RNS suddenly discover how patriotic the food-producers were for not getting any.

Secondly, the events of the crisis demonstrated very clearly how the safety-belt of guarantees thrown to the peasants in 1933 could change suddenly to a dead weight bearing them under. For Hitler's attitude was clear; once he had made concessions to restore profitability to agriculture, it must now defer to the interests of the community as a whole. Prior to 1933 the agrarian sector complained that it had been sacrificed to export interests. Now despite all the brave talk about a 'peasant policy' agriculture was going to take a back seat again, but this time economic stability was to have priority. Germany had an army of unemployed who had to be put back to work, mainly in industry, by means of public investment in labour-creation programmes. For these plans to be made fruitful the value of money had to be held as steady as possible so that public funds could be effective. Ultimately the Third Reich had to have a prices and incomes policy; the stability of one depended upon that of the other. Hitlerian fears of inflation simply combined with plain economic necessity to block any further price increases in food.

7

THE REICHSNÄHRSTAND IN ITS RELATIONSHIP TO OTHER OFFICIAL BODIES, AND ITS OWN LEADERSHIP CRISIS IN 1936-37

As the sole representative of German agriculture professionally the RNS was bound to come into contact, which turned out occasionally to be opposition, with both party and state. These two latter categories often overlapped in the Third Reich, so that the criterion used in selecting the heading for any particular issue discussed here is that of areas of authority. Where the existing administrative channels, at any level from local authority to national government, came into conflict with the RNS on a matter over which their machinery had normally exercised control prior to 1933, the question will be treated as one affecting the relations between RNS and state, whether the latter's representatives were party members or not. The grounds for bickering between the RNS and the normal administrative machinery were the question of mutual co-operation, RNS indifference to existing channels and, more importantly, the fields of competence in agrarian administration.

A lack of collaboration with the RNS became a cause of friction almost immediately after the formation of Darré's new organisation. As early as July 1934 local government officials were attacking it on this score, one even going so far as to describe the RNS as 'a state within a state' because of its cavalier attitude to the apparatus of governmental administration. The RNS clearly had a duty to keep officials apprised of its intentions but apparently was failing to do so. By November the *Regierungspräsident* in Hanover was complaining about a lack of co-operation which went right down the ladder of command to the lowest rung; in particular, no information about marketing arrangements, such as the formation of new *Verbände*, was being given. At the middle level of administration there was no liaison between the KBFs and the police, and new price regulations were not being communicated to local government officials. Similar comments came from Stade, where the RNS was described as 'dictatorial' and never ready to consult; the first that local authority in the district heard of new RNS measures was when the peasants came to complain about them. On the other hand, it was stated that at Hildesheim co-operation was proceeding wholly without friction, so that conditions clearly varied, individual personality on both sides being no doubt important.[1]

In general, however, the situation was unsatisfactory, which partly accounts for the November 1934 Conference already referred to (see page 79). Participants included the heads of local government in Prussia who were promised more information in future. Dr Saure of the Staff Office pointed out that there were now 514 KBFs in Germany, all relatively inexperienced in administration, but whose capacity would improve in time.

Similarly, efforts at a more successful partnership were also made at regional level; in Lower Saxony the LBF invited local authority representatives to a meeting to discuss the question of achieving more collaboration with the RNS in the area.[2] If subsequent reports are any reliable guide it would seem that the conference had been worth while, since the relationship then appeared to be distinctly better. This is not to say that friction never re-occurred; it was reported from Cologne in 1937 that the whole apparatus of state administration was threatened by subsidiary organisations, some attached to the RNS, some to the Four-Year Plan, whose existence led to confusion among the public, unable to discover who was responsible for what. The situation was generally unsatisfactory, in that opposition developed between the various bodies, each of which tried to shut itself off from the others.[3]

The creation of the RNS had produced problems for the existing political machinery, although these were more pronounced in 1934 than afterwards. Despite some improvement the essential point remained that by erecting a comprehensive organisation to deal with agriculture the NSDAP had produced a 'state within a state' which was bound to trespass on the domains hitherto reserved to local government. Even more indicative of private imperialism by the RNS were the efforts to take over some of its duties altogether.

The party began to assert the supremacy of Central Government over that of the regions in general almost as soon as acceding to power. In the area of agriculture in particular the Prussian Ministry of Agriculture was amalgamated with the REM, as from 1 January 1935:[4] by this means the friction that had existed between them under Weimar would be eliminated. As far as agrarian administration in the states was concerned, Darré had called it the worst of any developed country because of its lack of unity. Certainly there were great regional variations; Württemberg's rural parishes had spent RM 3.85 per head on aid to agriculture prior to 1933 whereas the national average was RM 0.63 only.[5] NSDAP desire for uniformity was probably in some measure a reaction; exactly how much store Darré laid on this can be gauged from an incident in 1935 when he charged Department I of the REM (under Willikens) with failure to get its policies put into practice especially in the south. The whole department was taken from him and handed over to Backe.[6] The takeover bids by the RNS for various regional authority functions must be seen against this background, and also within the general framework of a movement towards centralisation at the expense of the individual states over the whole governmental field in the Third Reich.

The initiative to replace the existing authority did not always come from the top of the RNS; in 1934 the LBF in Bavaria wrote to Darré suggesting a takeover of certain state duties in the area.[7] There was a similar onslaught in Württemberg by the LBF on viniculture and cattle-breeding which he tried to bring under his wing in 1934: negotiations broke down over finance. In February 1935 Willikens took the matter up again, basing his argument on the need to bring the south into line with Prussia, where the Agricultural Chambers had always performed some duties which the regional governments had done in the south: (this meant of course that they had come under the RNS in Prussia when that body absorbed the Chambers). The question was discussed at a meeting between the LBF and representatives of the Württemberg government in the following October. Eventually it was agreed in March 1937 that the RNS would now assume responsibility for cattle-breeding and viniculture in the area, receiving from the state government a sum equivalent to its own previous expenditure. With effect from July of the same year the state officials concerned became employees of the RNS.[8]

Apart from a natural desire to be masters in their own houses the regions in south Germany had other reasons for resisting takeover from the RNS, and using what delaying tactics were at hand to prevent it when they could. The costliness and allegedly bureaucratic methods of the new corporation were often cited as grounds for resistance. The administration by the RNS of forestry in Württemberg, for instance, was said to cost nearly three times the total which the Agricultural Chambers had spent for the same purpose.[9] Thus in Bavaria where the RNS assumed control of cattle-breeding institutes, considerable opposition was encountered. At a conference presided over by the Minister President Siebert (himself in the NSDAP) regional officials strongly attacked the RNS, the phrase 'complete failure' being used of its flour distribution in the area. To these criticisms the LBF Schuberth did not reply. Eventually, Siebert went to the lengths of commissioning three officials to draw up a memorandum on RNS inefficiency.[10] Clearly RNS pretensions were unwelcome to those existing channels which saw their own interests being threatened by Darré's relentless drive towards uniformity of administration on the land.

The vocational agrarian training schools were also a target in this respect: here Darré was occasionally unsuccessful. In Württemberg this started again with the LBF Arnold approaching the regional government and using the same approach as Willikens, by drawing attention to the fact that such institutions were under the Chambers (and therefore now the RNS) in most areas, and only under the region in Baden, Württemberg and Bavaria. Arnold suggested that professional training should be in the hands of the RNS in a corporative state. This claim was too large it seems, since the government decided to leave things as they were in the following month.[11] There was a long hiatus in the matter until 1938, when Rust, the national Minister of Education, produced a draft bill for the whole country. This put agricultural school supervision under his Ministry, whilst leaving the erection and maintenance of the buildings to the RNS everywhere. By 1942 no real agreement had yet been reached as the REM would not agree to the teachers being under another Ministry: equally it would not accept staff for the schools who were not in the RNS, which produced a deadlock.[12] It is interesting to notice that whether the Education Ministry or Darré ultimately won, the regional government was bound to lose. In fact in Hesse the RNS did manage to gain control of local training schools as early as October 1933.[13]

At lower levels of civil administration there was also considerable friction due to RNS poaching, for example, in regard to tenants on state farms; hitherto the domain of government officials, the latter were now not merely excluded, but not even told what was happening in the relations with the lessees. In one district even peasant complaints about the police were dealt with by the RNS. As already described, the peasants tended to treat their *Bauernführer* as Jacks of all trades, so that almost against its will the RNS became the vehicle for matters formerly the preserve of other bodies. So much did the local authorities become pushed into the background that dairy-owners in Göttingen were apparently surprised to be told that the supervision of their firms was still a government affair in any respect whatever.[14]

Such encroachment was usually produced by the conscious drive for uniformity in administration, sometimes because the RNS simply assumed authority for agriculture in all its aspects, which was bound to bring it willy-nilly into conflict with existing channels. For example, debt-relief was now organised via the KBF, which excluded the parish administration altogether. The best summary was perhaps the one from Osnabrück, that the RNS was a 'state within a state'. This has obviously some bearing upon the size of the new organisation and its cost to the members in the form of annual subscriptions; in assessing how high these were, it must be borne in mind that the RNS performed more duties than the previous professional bodies which it had replaced. In a certain sense it was the only genuine body in the Third Reich which could have been fitted easily into the framework of a corporative state; the *Reichsnährstand* was the estate of agriculture, in the medieval use of the word 'estate'. But in the Middle Ages, there was no sophisticated government apparatus with which such a corporation could conflict; transplanted to the modern era the 'estate' was bound to cut across lines of the existing sophisticated and complex network of civil administration.

This over-organisation needs to be contrasted with the lack of uniformity under the Republic. Darré seems undoubtedly to have been in favour of unity as such in agrarian administration and he over-reacted to the previous defects. Criticism of the RNS has therefore to be tempered with a sober assessment of the failings which he found on taking office. It was unfortunate for German agriculture to have swung so rapidly from one extreme to the other.

The RNS also indulged in some pretensions towards the political leadership of the peasants: inevitably some circles in the NSDAP did not take kindly to what was seen as an interference with the party as such, which saw purely political matters as very much its own affair. Two broad areas of conflict can be distinguished as the centres of the resulting struggle. First, there were those groups such as the DAF or the *NS Frauenschaft* which quarrelled with the RNS over areas of competence: this arose especially with the Gauleiters, whom Hitler had designated as his 'viceroys'. The donation of regional sovereignty to them almost inevitably provoked strife with the RNS as to who had the final say in agrarian matters in their respective fields. Secondly, there were those in the party, again including some Gauleiters, who simply did not accept the RNS as such. These people were far more serious opponents for Darré in that they wished to abolish the corporation rather than merely limiting its powers. The friction arising from contact between Darré and other leaders of the NSDAP was in some instances, however, occasioned by personal animosity; he had come to the party relatively late in comparison with the 'old fighters' of the movement, and his rapid rise to power seems to have provoked some degree of jealousy. But the main factor was RNS claims to precedence in political leadership on the land.

Conflict with the Gauleiters in general started quite early after Darré's initial appointment in 1930, when he came into collision with von Corswant, Gauleiter of Pomerania and an estate-owner, over the question of Polish labour. The Gauleiter wrote an article on the subject which he sent to Darré at Munich for approval, giving him eight days to read it; the latter, piqued, returned the article stamped 'not approved – Darré', and in a covering letter claimed that as a member of the Reichsleitung, he was superior to the Gauleiters, and would not dance to von Corswant's tune. The irritated Gauleiter replied that as an old member he refused to be dictated to by any new arrival anxious to display what he could do. Recriminations apparently terminated in a threat to report Darré to the party's tribunal of arbitration (Uschla): Darré reiterated that as leader of the agrarian section at headquarters he was superior to Gau organisations. Von Corswant then raised the issue with Uschla but did not get very far, as the latter body upheld Darré's precedence.[15] He had won an important first round against regional political leadership. Apart from that, the affair is interesting in that it illustrates Darré's tactlessness in dealing in such a fashion with a senior party member only five months after joining. Not for nothing did one of his contemporaries describe him as 'horribly vain'.[16]

Further trouble with the Gauleiters seems, however, to have been occasioned more by overlapping areas of competence than by personal failings and in November 1931 Strasser and Darré sent out a joint letter to all Gauleiters reminding them that LGFs were responsible for agrarian policy in toto as well as being advisers to the Gauleiter. In March and June 1932 Darré was obliged to repeat the delineation of frontiers of responsibility, as a result of further trouble. Then the Gauleiter in East Hanover, Telschow, told his staff that the nomination of *Bauernführer* required his counter-signature, which promptly drew an indignant letter from Darré to Hitler disputing the claim.[17] As far as the RNS was concerned the price of independent action appeared to be eternal vigilance.

In Pomerania even greater difficulties were experienced with von Corswant's successor, Karpenstein, a man with a pronounced dislike of the RNS. In a circular he described it as having cut itself off from the NSDAP altogether, and alleged that its officials refused to speak to the Gauleitung, and were in general acting as though the RNS was completely independent. Karpenstein told his staff that as he had seen no improvement in this respect, his patience was exhausted; if the RNS could not accept party leadership then its members should leave the NSDAP. The Gauleiter had tried to dissuade local *Bauernführer* from attending the meeting by threatening expulsion from the party for anyone who did so. Rumours had apparently been spread to the effect that the LBF's salary was RM 36,000 annually, which was announced at a meeting of OBFs by the political leadership: attempts were being made to subject the RNS press to party censorship, and the corporation was described in the Gau in general as bad, a pest and non-National Socialist. The war between party and RNS was evidently bitter here, although on the same grounds. The political leadership wished to make the corporation subject to its control, although of course in theory the RNS was an autonomous professional organisation, and not a branch of the NSDAP. How this particular battle in Pomerania would have ended is difficult to say, but in July 1934 Karpenstein was relieved of his post by Hitler. This probably had little to do with the friction between him and the RNS; the Führer's grounds for the dismissal were given as 'a series of strong complaints':[18] whether these included Darré's is possible, although clearly not decisive.

But of all the clashes between Gauleiter and RNS the most serious took place in East Prussia, partly due to the temperament of the local party chief, Erich Koch, a real old warrior of the NSDAP. He had apparently always ruled the province in a somewhat autocratic manner; the SA and SS had never been allowed to come to the fore, for example, as in other Gaue.[19] Even prior to 1933 he could not get on with the peasants, with whom he had little in common; his personal unpopularity became such that rumours about his possible exchange had to be officially denied. His lack of sympathy with both peasants and estate-owners lay in his belief in collectivisation as the best form for agriculture.[20] Thus a highly-explosive mixture existed in the shape of an autocratic party viceroy faced with the sudden formation of a new corporation in his domains, not always loath to claim independence; the outcome was a first-class row.

In July 1933 Koch gave a speech to officials from the party and the Agrarian Office in Königsberg; accounts of this differ considerably. The Gauleiter's version was that the NSDAP had not defeated thirty-six other parties in order to split into a similar number of internal factions, and that the task of the Agrarian Office was to turn the peasants into National Socialists. There would be only one line of development in East Prussia and any opponents could start thinking about a concentration camp; according to Koch his remarks received a mixed reception, but the LGF, Otto, walked out.[21] As the tenor of the Gauleiter's remarks was aimed quite clearly at RNS pretensions to independence, this was hardly surprising. Ten days later he delivered an open attack upon Koch: another member of the Agrarian Office named Witt went further and gave out a distorted version of the original Königsberg speech at a farmers meeting, alleging Koch to have said that anyone in East Prussia who talked about a free peasantry would be drowned in the Zehlau bog. The theme was taken up by a local farmer in the party. The Gauleiter had now allegedly said that if he could get the *Landesbauernschaft* under his thumb it would be an end to the free peasantry in the region. The fat was now in the fire and Koch began to arrest or expel his critics from the party, including members of the Agrarian Office such as Witt.

War was now declared between the corporation and the party in East Prussia; Witt wrote to Hitler protesting about the treatment he had received. Twelve days later Willikens took up the cudgels with a letter to Major Buch, head of Uschla at Munich, complaining of the Gauleiter's interference with agrarian self-government. Koch retaliated by arresting Otto as well; this drew Darré into the affair, and he wrote to Goering; when this letter came into Koch's hands he read it out in public, apparently with considerable misrepresentations. The sequel was a formal proposal by the RNS leader that the Gauleiter should be expelled from the party.[22]

However, not all the RNS officials in the province were necessarily opposed to Koch in his desire to keep the corporation under his direction. One LKF wrote to Otto in support of the Gauleiter, pointing out that to build up a professional organisation on the land without political supervision would be merely to construct another Green Front. This standpoint was apparently based on the assumption that the RNS was purely technical, although of course it was in reality dominated by NSDAP members, just as anxious to convert the peasants to National Socialism as was the political side of the party. But it would appear that even some RNS officials thought that political leadership should be left to the party.

By autumn moves in headquarters were under way to bring about some degree of reconciliation, and Uschla suggested to Koch that he should meet Darré himself; the latter then underlined his independence of the Gauleitung by nominating Otto and Witt as LBF and deputy, respectively, despite the row with Koch. The matter was now so out of hand that Munich forbade any public discussion of it to avoid possible damage to party prestige: as this came just after Darré's appointment of the LBF it would seem in retrospect that this latter move had really been rather a provocation on Darré's part; certainly Otto did nothing on his side to heal the breach, as he seemingly did not contact Koch in any way to discuss agrarian matters with him.[23] Hess now intervened as things had gone so far, and in late September Meinberg was despatched to East Prussia to try and settle the affair by personal contact; the subsequent report revealed that Gauleiter had instituted a virtual reign of terror in the province for the peasants. Incidentally, he refused even to see Meinberg at first.

In one district seven peasants, presumably hostile to Koch, had been arrested and remained in prison for a month; on their release they were compelled to sign an affidavit not to take action for their imprisonment. In some districts the *Kreisleiter* were said to be levying contributions from the peasants to pay for personal expenses; in Rosenberg two pounds of grain were taken daily from each farm to settle the local party's debts. In Heiligenbeil peasant contributions under the title of 'Winter Help and Work Creation' were subsidising both an NSDAP newspaper and providing a Mercedes for the party office: here the head of Department I of the RNS had been arrested on a false charge. A police official told Meinberg that the Gauleiter now scarcely dared to walk the streets of Königsberg in daylight because of feeling against him. Koch had also overstepped the limits of his jurisdiction in attempting to install a certain LKF as president of the local Agricultural Chamber, which the official had refused, as the Gauleiter had no competence in the matter: Koch promptly ordered him to resign all his offices or he would be sent to a concentration camp, adding 'For people like you there are special taxes'.

Clearly, local officials had been reduced to the belief that almost no one could work with the Gauleiter: the *Regierungspräsident* of Friedrich said that even if Otto were replaced his successor would soon experience the same problems: if on the other hand an LBF acceptable to Koch were installed the peasantry would be driven to desperation: one old party member was franker still, and said that either the Führer did not know what was happening or he did not care, in which case confidence in him would be lost.

With such a degree of antipathy to him from all shades of opinion it seems astonishing that the party chief could have survived at all, yet he did so, although eventually obliged to accept Meinberg as the LBF. The installation of Darré's deputy as head of the East Prussian RNS was presumably to ensure that Koch found himself faced with someone rather more important than the average LBF. In February 1934 there was a last rumble of the affair when Darré issued a decree on agrarian administration in which he described Koch's measures as having been harmful. The Gauleiter promptly protested against the tone of these observations, and asked Darré to see that his subordinates did not speak of 'old fighters' for the party in this way.[24]

There were several factors to consider in assessing the whole row. Koch's collectivist attitude towards agriculture was important, as was his personality, which was stormy and capricious; thirdly, there was again a certain tendency to distrust Darré as a newcomer. The final issue is the most important, the role of the RNS in relation to the party; the Gauleiter told Meinberg that he had nothing against the corporation as such, only 'some pests' in it. A review of his activities does not support this statement: Koch was clearly a man who took the Hitlerian injunction about vice-royalty literally. Under these circumstances he was almost certain to conflict with the RNS over decisions about areas of authority. Like Telschow and Karpenstein, Koch felt agrarian administration to be a party matter. Ultimately, the blame for the friction must in all cases be laid at the door of Adolf Hitler, who by accident or design had created geographical empires for the Gauleiters and a vocational one for Darré. Clashes could then scarcely be avoided.

As a result of repeated conflict the party leadership saw itself obliged to intervene, Hess being the arbitrator in this instance. He called a general Gauleiters' conference on 13 December 1934, to discuss the question of party/RNS relations. Two months later a communiqué was issued. This stated that as the RNS had received a commission from the Führer to guarantee German food supplies, the party was henceforth forbidden to intervene in the leadership of agriculture at a purely technical level. All future appointments of *Bauernführer* would be made in consultation with the appropriate Gauleiter. Hess referred to the question of women in rural areas, which had also been the subject of friction, in this case between the RNS and the *NS Frauenschaft,* the party organisation for women.[25] In future, the RNS would use its influence to enrol rural housewives in the party but would cease to have its own office for them. Party experts would be appointed for this sphere, although they would only act in consultation with the RBF. Hess's announcement finished by saying that all public meetings and demonstrations for peasants were henceforth to be a party matter, except where the subject matter was wholly technical.

He then expressed the wish that close co-operation would be possible in future between the party and the RNS, with the political primacy of the first being maintained. This was the keynote of his announcement; true, the Gauleiters could not meddle in purely professional questions, but these, judging from the evidence of conflict, had never been the cause of the strife in the first place. The real issue, who was the leader in the political sphere, had been resolved wholly in favour of the party and its regional representatives, the Gauleiters. A firm limit had been put to Darré's private imperium, which was now to be seen as a professional corporation under the political aegis of the NSDAP.

The intervention by the Führer's deputy did not by any means settle the tension between party and RNS, although it is true that a temporary peace was patched up. But that trouble continued to exist may be inferred from Darré's allegations in 1938 to the effect that two different Gauleiters, Bürckel in the Saarland and Lohse in Schleswig-Holstein, were both trying to set up new agrarian-political branches in their organisation, independent of the RNS.[26] Despite Hess's ruling some Gauleiters always cherished hopes of getting agricultural leadership totally under their own wing.

This applied equally to party headquarters as well. Darré's original agrarian office at Munich was virtually inoperative after the NSDAP accession, since its members joined the RNS anyway and became its leaders. In 1937, however, efforts were made at the party chancellery to build up an independent office for agrarian-political matters on the land. Darré therefore sent Motz of the RNS to Munich to reactivate his old agrarian apparatus in order to take the wind out of the party sails.[27] The incident is typical of the internecine warfare waged at various levels by Darré in order to preserve the integrity of his private empire.

In his announcement Hess had spoken of the landworker question, which also presented a thorny problem in terms of overlapping authority, as the RNS was the corporation for agriculture but Dr Ley's Labour Front (DAF) had the monopoly of the administration of labour. Who then administered farm labour? Yet another struggle took place in the corridors of the Third Reich on this issue.

This was important, if only because of the number of persons who formed the object of the disputes, and their low social status, which the NSDAP inherited as a problem. The two most superior grades were the *Deputate* who had a piece of land as well as wages and received some payment in kind, and those classified as *Heuerling,* who also held land but virtually under feudal terms, in that they worked so many days annually for the owner, without however being his direct wage-employee, like the *Deputate.* At the bottom of the scale came the day labourers *(Tagesarbeiter)* and the farm servants living in *(Gesinde).*

Conditions varied therefore for farm labour both in terms of occupation and by region: the two distinctions were sometimes really only one, as the different soil conditions in the various regions determined which type prevailed. The *Heuerling* was found most often in the north-west, the day-labourers on eastern estates (where in 1933 it was still common practice for them to kiss the hem of their employer's coat when greeting him).[28] Even more revealing of the labourer's status in Prussia were the results of Weimar legislation. The Prussian government had instituted legal procedures against ill-treatment of farmworkers, to enable them to obtain redress: between October 1928 and February 1930 forty-six cases of physical mishandling of labourers from East

Prussia and Silesia alone were reported as a result of this decree.[29]

The NSDAP clearly had a sizeable task on its hands if it wished to improve labour relations on the land, as it undoubtedly did. The bedrock of its outlook was simply that the standing of farm labour should be improved in such a way that the class-struggle on the land would be overcome by rubbing out the line between employer and hired hand to some extent, not by lowering the status of the former but rather by improving the position of the latter.

The struggle between RNS and DAF must be seen in this framework, as one concerned with methods rather than with goals, as both corporations agreed that the landworkers should be raised to the status of *Heuerling* wherever possible, which one *Landesbauernschaft* described as 'the soundest form in agriculture':[30] having land of his own was good training in management for the farm labourer, acting as a kind of springboard which the vigorous could employ as the first step to becoming *Bauern* in their own right.[31] The DAF similarly regarded *Heuerling's* status as the best goal; one article in *Arbeitertum* recommended that every labourer should get two to five Ha of land in order to overcome Marxist conceptions, as witness the settlement company in East Havelland which gave plots of five Ha each to landworkers.[32] In sum, the objective of the NSDAP was to remove as far as possible the distinction between employer and employed in order to overcome the class-struggle (and combat rural migration).

There was agreement over objectives, but the approach was different. The DAF almost inevitably as a labour corporation started from the workers' standpoint, whereas the RNS favoured the peasant employer. The occasions for the clashes were usually the living-conditions of farm labour, and its social status. Admittedly, the RNS was active in both these areas: a system of using particular workers as labour representatives was built up in each locality. These were appointed by the KBF to be responsible for the 'legal and social importance of his fellow-workers in the district'. There was apparently more than one in each village, as one list gave fifty-three names for twenty-five localities.[33] The KBFs themselves were always ready to deal with bad conditions for labour; after a visit to one holding a KBF in Lower Saxony wrote to the employer concerned that he should provide a separate bed for each labourer and improve accommodation in general, but this was a request rather than an order. An estate-owner in the Rhineland was told that insufficient bread was being provided at breakfast for his workers, that working hours were too long and that the manner of his agents towards the labourers was 'unsocial'.[34]

In other words Darré's corporation made genuine efforts to improve conditions in general for the landworker. However, the sheer scale of the problem defeated them in most cases; not the least of the difficulties which the RNS had to overcome was the traditional peasant attitude to hired labour, which was often backward in the extreme. Many employers regarded their farmworkers simply as 'beasts of burden'.[35]

This might not have made any difference to party harmony had not a law been passed in 1934 to regulate wages and conditions of service for all labour. A potential troublemaker was the provisions regarding conditions of service, which henceforth permitted the DAF to intervene everywhere, including agriculture, so that employers and labour would work as a team 'for the common good of people and state' and overcome the class-struggle.

There is every reason to suppose that Darré agreed with such an aspiration, but the new law created difficulties for him by establishing special Courts of Honour where workers with a grievance could formally seek redress. This was equally valid for farmworkers and they took full advantage: in 1935 nearly one-quarter of all recorded cases were in respect of agriculture.[36] A typical instance concerned a peasant who had failed to treat a day-labourer correctly; when reprimanded by the DAF he dismissed him, was hauled before the Court of Honour and fined RM 1,500. Here the possible clash of competence between the DAF and Darré's corporation was really brought into the open, since after all the RNS was being supplanted in a case where it had failed to take action. In another later case a peasant was convicted of providing totally unsatisfactory accommodation, and was deprived by the Court of his title as 'management leader' (which description applied to all employers under the 1934 law) and his son was fined RM 100. Moreover, in other areas too the DAF was not backward in intervening in rural matters; the relief of overwork by farmers' wives was held to be a duty of its female section.[37] Again, this kind of claim could only be interpreted by the RNS as a slur upon its own efforts.

Something had to be done to define the limits of responsibility more exactly, and in 1935 Dr Ley and Darré managed to reach a settlement: known as the Bückeberger Agreement, it delineated the frontiers between DAF and RNS in five clauses.[38] Under these, the RNS officially became part of Dr Ley's organisation in principle, replacing the former Section 14 of the DAF, hitherto responsible for agricultural labour: RNS members were to enjoy DAF facilities, such as 'strength through joy' holidays, and would then contribute financially to Ley's organisation. RNS leaders would automatically get places in the internal governing structure of the other body, and the RBF would appoint his advisers in socio-political matters in agreement with the DAF. In theory, this looked like a solidly-based accord but in practice the RNS subordination to the DAF was a form of words only. Moreover, if landworkers paid a subscription to the DAF they would naturally utilise it, in which respect the agreement was a dead letter almost at once. This had the corollary of arousing the ire of agrarian employers who evidently did not want their workers in a different corporation to themselves, perhaps because they felt that the RNS was more 'peasant-orientated' than the DAF; in one case a maid and a labourer were dismissed by the farmer when they paid DAF subscriptions.

Friction between the RNS and the DAF recommenced despite the Bückeberger Agreement, for example, between Ley's district office in Niederberg and the local KBF.[39] More serious trouble arose when in February 1937 the Ministry of Labour attempted to get provision for DAF representatives to be appointed even in enterprises employing between five and nineteen people, in other words, to quite small businesses. This would apparently have been applicable even on the land, since the REM protested vigorously and action on the point was postponed. Darré had won at least one victory against the pretensions of the DAF.[40] Ley's corporation was by no means in a mood to surrender, however, and a few months later was allegedly trying to keep RNS representatives for farm labour out of office altogether in Brandenburg. By now even farmworkers' wages had become an issue, since in Saxony the KBFs were demanding that the local DAF organisation should stop its propaganda about low pay for agricultural employees, which simply added to their dissatisfaction.[41]

By 1938 Darré's campaign reached its zenith, as the DAF encroached on him in one area after another. He alleged first to Goering and Hitler that Ley had said openly in January that he wanted to dissolve the RNS or take it over completely; under such proposals the land would be subject to the city, in effect. Darré did not miss the chance of complaining about the DAF's handling of the farmworker question as well.[42] Later in the year he took up this question even more energetically with Hess and indicted Dr Ley's organisation on three counts. These were the attempts to take over enterprises already affiliated to the RNS, the enrolment of individual members in a similar way, and the DAF concentration upon the living conditions of the labourers, which overlooked how badly off the peasants were as well. In support of his thesis Darré produced evidence of DAF interference from Munich, Dortmund, Schleswig-Holstein, Thuringia, Saxon-Anhalt and Pomerania – a formidable list.

In particular he found it irritating that the impression was now being given in party circles that nothing had been done for farmworkers until the DAF came along, in which propaganda the party journal *Der Angriff* was involved. Number 54 described how only the DAF had the landworkers' confidence, whilst a later issue described rural living conditions under the title 'In the land of the poor people': this told of the pleasure brought into landworkers' lives by 'strength through joy' holidays in the Rhön Valley. Darré pointed out to Hess that the RNS had provided many similar facilities itself (which no doubt explained his pique). He finished by attaching a seven-point draft to delineate the frontiers of responsibility between himself and Dr Ley, and demanded that the RNS be given greater support by the party to enable it to carry out its duties. His complaints were laid before Hitler who eventually decided to re-define the whole position of the DAF three months later.

Dr Ley himself did reply at once to Darré in writing, in a way that rubbed salt into the wound: the DAF had only intervened on those occasions when the RNS had appeared to be inadequate. He drew attention to farm sales, which had increased by nearly RM 2.5 billion between 1933 and 1937 whilst agricultural workers' wages had gone up by only about one-tenth of that amount. The whole problem of the landworker, he suggested, required an integrated approach by the party and its agent, the DAF.

By September 1938 Darré was bitterly complaining once more. The RBF enclosed examples of encroachment from twenty-five different places in the Reich at this time. He quoted one RNS official from Wismar who had said that the rural population itself could not understand the apparent double administration of both his organisation and the DAF: he referred to 'over-organisation' for agriculture as a whole. Similarly, from Landau it was reported that the DAF had taken on the administration of Italian landworkers as though the RNS did not exist.

Despite Hitler's previous intention to define the DAF areas more exactly, nothing much happened and by January Darré was vocal yet again. The whole root of the matter was labour relations on the land. The DAF spoke of the difficulties inherent in getting labour disputes settled as the farmers simply refused to accept Labour Front authority on the grounds that they belonged to the RNS. Darré refuted indignantly any suggestion that the RNS could not deal with such matters off its own bat and quoted figures to suggest that his officials were more adept at it. The Minister felt that the whole stress of the DAF was upon employers' failures on the land, a tendency which he designated as Marxist: leadership of people on the land could not be separated from farm management (presumably this was intended to imply that the RNS was better able to deal with the matter by way of its combined approach).[43]

It is clear that the dissension between Dr Ley and Darré was characteristic of the Third Reich where private empires so often collided. It may well have been that Hitler intended this, on the principle that if Darré quarrelled with the Gauleiters or with Ley then he would not have time to conspire against the Führer himself. This principle of leaving subordinates to attack one another had the added advantage of making Hitler a kind of one-man Appeal Court, to whom Darré and others frequently had recourse in order to obtain a definitive ruling on some matter or other. Hitler's edicts on such bodies as the RNS, or the DAF, or the position of the Gauleiters, created such an obvious clash of interest in terms of administration or political power that it becomes difficult to imaginc that this was not clear to him at the time.

Not all party members even wished the RNS to exist at all. There was a hint of this attitude in Ley's comment to Darré that the landworker problem could only be solved by an integrated approach by the party and the DAF. Behind such statements was the whole issue of the corporative state as such. The rights and wrongs of this concept cannot be discussed here in principle: what is pertinent are the conflicting beliefs in the NSDAP itself on the question. In theory, the movement accepted corporations such as the RNS, namely, autonomous bodies based on trade or occupation. As one contemporary account put it: 'In principle, the National Socialist state strives for a political structure based on estates'.[44] Against this, the party spokesman on the corporative state wrote that the NSDAP desired these structures only in an economic sense 'The National Socialist professional build-up does not have the state as an object'.[45] To add to the confusion, there was a third school of thought opposed to the corporative idea in any shape or form, which was the view of many leading party members.[46] To judge from Hitler's own remarks to the Reichstag in March 1933, he may well have been in the latter group, since he went out of his way to draw attention to his opinion that German economic recovery should be left to private enterprise, and that no elaborate bureaucratic mechanisms would be erected to solve the problem: the Führer was no economic theorist. In sum, there was total confusion in the ranks; as Ley said, he had never met two party members with the same views on the corporative state and its organisation, and Rauschning averred that in some party circles the whole idea was anathema even in principle.[47]

A great deal of the opposition to the RNS already cited here must be seen against this background. There simply never was any clear agreement about what form the National Socialist state should take; this reflected the movement's unique composition and background. Unlike many modern political parties the NSDAP was not founded on economic theories. It was rather a collection of individuals opposed to liberalism and Marxism, and whose overriding principle was race. All else was pragmatism and tactical manoeuvre.

In early 1937 the whole question of the RNS's place in Germany was thrown into sharp relief. Darré was away ill in the latter half of the previous year and Meinberg was left to carry on in his absence. His general conviction appeared to be that the corporation existed for the sake of agriculture alone, rather than for the wider community.[48] Consequently, he was reluctant to take orders from governmental circles, notably from Backe as Secretary of State in the REM, with whom he waged a continual battle, which Darré uncovered on his return to duty.[49] The Minister himself was firmly on Backe's side, since 'in the authoritarian NSDAP state' governmental channels took precedence over the estate, as the former represented the whole people.

In addition the deputy RBF had apparently also been guilty of personal transgressions, such as promoting his own favourites and making 'derogatory utterances' about the party political leadership. As a result of Darré's subsequent investigations a complete shake-up of the whole organisation was carried out. Meinberg himself had to go, as well as two departmental leaders and other leading officials. Darré gave his deputy the choice of dismissal, or resignation. The Minister himself preferred the second course, as had Meinberg been openly dismissed opponents of the RNS would have had, as he said, a chance to attack it with redoubled strength. Here again Darré displayed a certain degree of sensitivity to the general unpopularity of his own corporation and how eager critics were for ammunition against it. Eventually, after a few more rumblings from Meinberg, including the outburst 'If I go to him [Darré] it'll be with a dog-whip', he resigned, but only after considerable pressure from the top; the final offer from Hitler, as relayed by Goering, was a new post. In exchange for his retirement from the RNS the disciplinary process against him was to be broken off. To preserve the façade of unity so dear to the Third Reich he was ostensibly given control of credit and investment under the Four Year Plan for agriculture; his personal file shows that in fact he merely went to the Herman Goering Works. Gustav Behrens became the new deputy RBF, having formerly been chairman of a commodity federation.

The affair shook the corporation badly, and seriously damaged its prestige; Darré claimed that it had required 'an almost superhuman effort of will' to regain the confidence of the Führer and Goering. Needless to say, news of the matter spread and gave opponents of the RNS the chance to try and undermine it. In Essen Gauleiter Terboven told the Rhineland LBF that Darré's organisation was waging war on two fronts, against the party and the rest of the economy. He suggested that the NSDAP motto should now become 'Away from the corporative principle and back to the party and the DAF' (cf. Dr Ley). Terboven finished by asking the LBF if he really thought that the RNS was the best vehicle for the peasantry; the latter predicted that the Gauleiter would soon be at Goering's ear with this viewpoint.[50]

Unhappily for the corporation yet another row broke out almost simultaneously, centring around Habbes, the deputy LBF in Westphalia.[51] A Gestapo report in April 1937 put Darré on the alert and the deputy LBF was suspended pending an investigation. Eventually Habbes was accused of personal failings, but like Meinberg, he had apparently been trying rather too strongly to represent the peasants' interests against the state leadership. Again this brought the corporation into dispute, both with the party and with the rural population in Westphalia. The latter showed its disapproval in a concrete way, by reducing donations to the Winter Help scheme.

Dissension with the local party was occasioned partly by administrative organisation; the regional *Landesbauernschaft* cut geographically across both the Gaue for north and south Westphalia. Unhappily the respective Gauleiters had differing opinions regarding the capacity of Habbes. As the court of inquiry was unable to prove that he had intended any detriment to the party he began to demand reinstatement quite energetically. The embarrassed Darré provisionally allowed this and then ran into fire from north Westphalia for having done so, having previously been attacked by the southern Gau for removing Habbes in the first place. Eventually a compromise was reached.

Once again, apart from causing party/RNS friction, differing views about the status and duties of the RNS had occasioned discord in its own ranks. The temporal coincidence of the rows over Meinberg and Habbes must have reinforced the conviction in the top political leadership that there was something rotten in Darré's corporation, and undermined Hitler's confidence in his subordinate. It is also an interesting commentary on the RNS's claims to be self-governing that anyone who took this point seriously soon found himself in trouble. Under the pressure of the dismissal of Meinberg and the needs of the Four Year Plan, the *Reichsnährstand* lost any sort of independence which it may ever have enjoyed and simply became an outright instrument for the mobilisation of the agrarian sector and its adaptation to a wartime economy. Indeed by 1939 the introduction without friction of a wartime food system was advanced as the main justification of its whole existence.[52]

8

THE ERBHOF LAW

The Background to its Introduction, its Provisions and Reception upon the Land

The NSDAP's Agrarian Office decided quite soon on inheritance reforms after accession to power, in order to ensure that in Prussia at least there would in future be only one eligible heir for each holding. This was common practice already in north and east Germany, but as promised in the March 1930 programme this was to become a legal obligation. In early 1933 only Prussia could be considered for new legislation as the Reich Minister of Justice was not a party member, whereas in Prussia Kerrl of the NSDAP was Minister of Justice.

By April discussions were in full swing over a reform, the prime movers being Backe, Willikens and Kerrl himself: on 15 April Backe sent an interim report on the proceedings to Darré for his information.[1] There were several points on which agreement had not yet been reached: in particular, there was the question as to whether inheritance should be closed in the male line exclusively, as well as the problem of what size range of farms should be comprised in the legislation; finally, there was the very difficult issue of compensation for those sons and daughters who would now become barred from receiving actual land.

The normal practice in districts with closed inheritance (*Anerbenrecht*) had been to give them monetary payments in lieu, a custom known as *Abfindung:* this had in recent years led to considerable indebtedness, however, as low prices meant limited incomes for farmers, so that debts had been incurred because *Abfindung* could no longer be paid from current receipts, as independent sources confirmed.[2] This did not apply to districts with partible inheritance (*Realteilung*) as all family members there received land anyway, which was why the indebtedness in the Rhineland was the lowest in Germany.[3] The National Socialists could not accept divided inheritance, however, as it led in the long run to fragmentation of the property, and was traditional in limited areas of the country only. As this was excluded then the only choice remaining clearly seemed to be *Anerbenrecht* without *Abfindung.* On this point Kerrl was strict; he took the line that children other than the heir should be entitled to receive vocational training only at the heir's expense, but no lump sums in lieu of land, as had been the custom; daughters should be entitled to a trousseau, but not to a dowry. To the farming community, accustomed to giving monetary payments to all family members other than the sole heir, this was a sharp departure from recognised practice, but it has to be accepted that the logic of mounting indebtedness seemed to imply radical measures.

One issue not discussed in April was in some respects the most important of all, whether the new law should be comprehensive. Such a measure would mean protecting all the peasantry against speculation, on which Backe himself had reservations, as this would entail sheltering the 'less worthy' as well as the others. However, the Act as passed allowed the peasant no choice and thus included not merely the racially sound, but also the biologically 'less worthy'. This is curious, as by the NSDAP's own standards the latter could hardly have been considered suitable as the 'life-source' of the nation.

Darré was able to inspect the draft before it went to Hitler, interestingly enough, as he was at the time in no way a member of the government. A Civil Servant in the Ministry of Justice, Gustav Wagemann, also played some part in framing the legislation. Kerrl seems to have been particularly influential, and later referred publicly to the law as 'my finest work'. It is interesting that it was later said that the old *Anerbenrecht* law in Hanover was Kerrl's inspiration, as he came from Lower Saxony.[4] As Hanover did have a compulsory law of this nature until its abandonment in 1866, there seems no reason to doubt the truth of this statement.[5]

The Act was published on 15 May 1933, and comprehended all peasants in Prussia owning holdings of between 7.5 and 125 Ha in size: these were to be henceforth known as an *Erbhof* (plural *Erbhöfe*), and only their proprietors would merit the honorific *Bauer.* (From now on this word designates here the owner of an *Erbhof* only.) No such holding could be sold without court permission, nor could foreclosure be applied against it. The farm was thus strictly entailed since the *Bauer* could name only a sole heir from among his children, and the farm would be protected against the play of free economic forces. If sale became unavoidable, the members of the family had priority of purchase. To qualify as an *Erbhof* the holding had to be in the possession of a sole owner, those jointly owned being ineligible.

Abfindung could be now paid from current receipts only – Kerrl had had his way. No debt or mortgage could be secured against the holding in this respect. The family farm became thus even further removed from speculation, the elimination of which was the main object of the law. The new measure has to be read in conjunction with the almost simultaneous introduction of guaranteed prices, as these clearly affected *Abfindung.* The provision allowing this from current receipts only makes sense, and appears less harsh, when it is realised that the RNS envisaged that from now on farm incomes would be far higher anyway.

Almost as soon as the law for Prussia had been published the NSDAP began to put on pressure, either openly or behind the scenes, to get it extended to the Reich. Obviously Darré's own assumption of the Ministry of Agriculture was a major factor in facilitating this. He called publicly for the extension of the legislation nationally in order to preserve the peasantry as the biological fundament of the people, the usual NSDAP slogan. In the regions there was similar activity; in Württemberg Arnold urged adoption of the law at a government conference, local opinion being on the whole welcoming but reserved.[6] Pressure was brought to a successful climax with thc publication of the *Erbhofgesetz* (EHG) itself, which in effect extended the Prussian law to the whole Reich in September 1933.

The new legislation was a lengthy and complicated affair. Several articles dealt with the two concepts *Bauer* and *Erbhof,* which were defined as in the Prussian law except that the previous specific size range of 7.5 to 125 Ha was abandoned. The holding had now to be large enough to support the whole family so that they required no other sources of revenue in order to live. Local conditions were therefore to be the determining factor, as was laid down by the highest court that 'the question of what quantity of land suffices to make a family independent . . . has not been decided by the law'.[7] This meant the local court had to determine whether the holding concerned was viable or not in view of conditions on the spot; at the same time it did at least ensure that these latter circumstances would be the criteria, instead of the fixed size ranges laid down by the measure in Prussia. This move towards a greater degree of flexibility may well have been in response to regional demands, and therefore represented perhaps a *quid pro quo* in return for acceptance of the measure nationally. Certainly both in Westphalia and in Württemberg the initial response to the Prussian law had been an expressed desire that local conditions should prevail in the interpretation of any general Act.[8]

The *Bauer* had to be of 'German or similar blood' which excluded Jews and coloured people. 'Similar blood' was defined very widely in the outcome, embracing all races settled in Europe since the start of recorded history.[9] The exclusion of Jews did not have much practical consequence, as few were farmers anyway. Moreover, they could still own land as such; in 1936 there were eighteen Jewish-owned farms in the Rhineland, plus half that number again of parcels of land used by cattle-dealers for grazing, but these were by definition ineligible to become entailed farms.[10] The *Bauer* was compelled to prove his Germanic or similar descent since 1 January 1800 in order to qualify for the title, probably because of the dissolution of the ghettos in Germany between 1797 and 1811. Before that time Jews and Germans simply had not mixed. The *Bauer* was required at first to be of German nationality although a later amendment allowed the REM to dispense with this condition where it saw fit.[11]

The peasant had to be as before the sole possessor of the property, holdings jointly owned not being eligible for enrolment: he similarly had to be both efficient at his profession, and honourable, a special court being designated as the arbiter of these qualities in cases of doubt. In the event of his losing either or both, the wife or heir could be designated as farm manager; if neither were available the RBF could name a new occupier. Under such circumstances the dispossessed peasant (no longer a *Bauer*) could lay any financial claims outstanding on the property under the Civil Code of 1900. Loss of the title *Bauer* and dispossession made no difference to the holding, which simply remained an *Erbhof* under new management. The whole procedure was typical of the NSDAP slogan 'The common good before individual gain', and the limitations on private property rights which the party had always demanded for agriculture. As Hitler put it at the Harvest Festival in 1933, 'the NSDAP will preserve the race even if necessary at the expense of the individual'. The new law mirrored the monism of the party's approach to society in general. The individual *Bauer* might have to go but the family would remain. As the preamble emphasised, 'Blood and Soil' was now a reality.

Article XX laid down the order of precedence in the nomination of the heir. This ran: (a) testator's son, (b) his father, (c) his brother, (d) daughter of the testator, (e) sisters, (f) any other female descendants. This very open preference for males contradicted normal German practice. In order to soften the blow a special arrangement was allowed as a kind of transitional period; where no son was available when the law came into effect (1 October 1933), daughters could take precedence over the father or brothers. When any daughter so inheriting came, however, to dispose of the property in her turn, this concession would automatically lapse (Article XXI). The Act, incidentally, did not establish primogeniture, merely closed inheritance. Whichever son the *Bauer* chose to designate as heir should be decided according to local custom. Where there was none, the youngest son would take precedence. The thinking behind the priority for the youngest was that the older sons would have more chance to make a career for themselves if they did not wish to be farmers.[12]

A further article softened the effect of testamentary restrictions by giving concessions over the nomination. In those districts where *Anerbenrecht* had not formerly applied the *Bauer* had a free choice. This also applied to regions where 'important grounds' for deviation from the legally-determined line of precedence could be shown; this could even include advancing the claims of the daughters, but the onus was on the *Bauer* to show why it was justified. In those districts where free choice of heir had been the rule this could still be exercised. The testator could also (Article XXVI) authorise the father or mother of the heir to manage the farm until the latter attained 25 years of age. This allowed the widow to exercise control over the farm in some cases, if only for a limited period, but she could never inherit the property, which had to remain in the same blood-line. This was not perhaps as harsh as it might seem, since the normal practice was to bequeath the holding to an offspring. The latter normally took over the farm when the parents reached 65 to 70 years of age, a contract being signed which allowed them to live on the farm at the heir's expense, usually with stipulated payments in cash or in kind. When seen against this background the law's provisions regarding inheritance appear less restrictive than a cursory reading of them might indicate, especially in respect of those regions (the majority) where *Anerbenrecht* was the established pattern already.

The designated heir was not obliged to accept the holding because under Article XXIX he was entitled to forswear his claim within a period of six months of being told of the owner's intentions. A court at Wuppertal-Elberfeld accepted an elder son as heir because 'the younger son . . . has renounced his right of inheritance'. In one case in Württemberg the local court allowed an elder son to take over in place of his brother who did not wish to work the holding despite his nomination.[13] There was apparently no pressure upon the younger people to accept the farm if they did not feel like it.

Article XXX established a curious right known as *Heimatzuflucht*, which provided family members not inheriting land the privilege of board and accommodation at the family holding at a time of financial need. This must be seen in the framework of the new regulations which confined their aid merely to vocational training at the heir's expense. As the old custom of granting them direct monetary payments was now discontinued they received the education and *Heimatzuflucht* as compensation. Apparently, the latter right was not normally invoked because Germany was undergoing such an economic boom that its provisions simply did not apply. Had there been a new depression, however, the peasantry might well have been submerged under a flood of indigent relatives. In other words, the fact that it was not used shows how much better off rural people were if they migrated to the cities.

The same section dealt with the entailment of the holding; neither as an entity nor in part could it now be sold without court permission. Debts had henceforth to be payable from current receipts, and no mortgage could be taken up against the holding as collateral without special permission. This corresponded to the party programme of March 1930. So strong was the NSDAP conviction that the peasantry should not be burdened by debt that in 1936 a further decree came out disqualifying any holding from becoming an *Erbhof* where its current debts were over 70 percent of the total farm value at the last estimate.[14] Security against foreclosure was an essential provision: Article XXXII did, however, allow a creditor to take farm produce provided it was not required for family consumption by the *Bauer*. Land could not be proceeded against by a creditor under any circumstances whatever.

Wilhelm Meinberg was later to claim that the EHG had lifted the

Bauer from out of 'capitalist debt-servitude'. Clearly there was some substance in this statement as virtually total security of tenure was now given to the family. As against this, the individual farmer could be removed from management for personal failings. The NSDAP would doubtless have accepted this, but replied that under the capitalist system he could have lost the holding by compulsory auction, which was now forbidden. One weapon against the *Bauer* was in this respect exchanged for another, but the real difference now was that whatever happened, the holding stayed in the same family.

Even in principle, the EHG entailed rather less restrictions than might at first sight appear. Although closed inheritance was obligatory, most districts had this as a custom already: had they not, the clamour for a national law could not have arisen in the twenties. Moreover, the clauses in the EHG regulating the hereditary succession were hedged around with all sorts of modifications, especially for the transitional period following the law's introduction. In particular, local custom was always to prevail in deciding which particular member of the family should inherit: in this respect, the feelings expressed in such regions as Westphalia and Württemberg, that the EHG should be a set of general principles rather than a blueprint to be followed in the same way, everywhere had been met. The open admission of local attitudes had perhaps been of some influence in shaping the EHG in this way. In the case of *Abfindung,* however, this was not so, as it was now rigorously excluded everywhere. Here the NSDAP had yielded no ground, in view of its declared war on farm indebtedness. Henceforth all such payments could be from current receipts only.

Two further points have to be made regarding the EHG in general: first its effect in *Realteilung* districts. Here the obligatory nature of the legislation clearly made a greater impact than in the rest of the country, but some qualification is necessary even here. The very fact that *Realteilung* had led to property fragmentation meant that fewer holdings in these areas were eligible as *Erbhöfe* in the first place. Whereas in Westphalia roughly half the agricultural land was eventually comprehended under the EHG, in Baden, the Rhineland province of Prussia and Hessen-Nassau the relevant percentages were 16.6, 15.6 and 19.4, respectively.[15] As *Realteilung* had been widespread in the latter areas a smaller proportion of the peasants was touched by the new legislation.

Secondly, unlike the marketing regulations, or the RNS in general, the EHG applied to peasant holdings only, which represented about 70 percent of all those of 5 Ha or above in the country: the REM estimated that potentially there were 1,071,300 *Erbhöfe*. It was the *Bauer* and not just all farmers who became specially privileged, although the latter did enjoy the benefits of guaranteed prices, of course; but only the *Bauer* could not be evicted from his land for debts. A hint of this attitude had been given by Darré even before the NSDAP's accession when he stated that a racial state could not be built up on an agriculture based on farmers (*Landwirte*). Ideologically it was the peasant with his traditional attachment to the soil of his ancestors who had to be the foundation of the people. Hence the ban on speculation with the farm; once the owner did that he ceased by definition to be a peasant anyway, as he was allowing economic considerations to prevail over 'blood and soil'.[16]

However, in some unusual cases farmers were allowed to enrol as *Erbhofbauer,* but only when they could prove that the land had been in their possession for over 150 years. By mid-1938 some 1,000 farms over 125 Ha in size had been registered in this way; as this represented less than 4 percent of holdings of this size, clearly such cases were exceptional.[17]

As well as guarding the *Bauer* against future indebtedness, strenuous efforts were made to relieve him of existing liabilities, if unduly heavy. This caused a certain amount of controversy in high places, the dispute being as usual on the question of just how comprehensive such legislation ought to be. Should all *Erbhof* owners be included in the scheme or not? At the Reichsbank, the citadel of financial orthodoxy, it was felt that some kind of discrimination was necessary. It was suggested that honourable *Bauern* would not be in debt;[18] this was turning the NSDAP's agrarian ideology against the party itself, in effect. When the matter came to a general discussion there was a lively debate, in which the Finance Ministry representatives held out for relatively high interest rates to be charged to the *Bauer* as his share of paying off his commitments.

Eventually a compromise was attained under which the *Bauer* registered all debts as at 30 January 1933 with his LBF provided that the property as a whole was worth RM 5,000 or more. The state took over all these debts as from 1 October 1933 under the slogan 'The *Bauer* and his farm will be free'. In return the indebted peasant paid 1 percent of his total farm value yearly to the government and a further 2 percent on the debts taken over. Some criticisms of the new arrangements came from von Rohr, who alleged them to be less favourable than Hugenberg's debt-relief act.[19] He suggested that 40 percent of the indebted owners would now have to pay more. His figures appear slightly distorted, however, by being based on an example of a holding worth well above the average. Other commentators found the new scheme to be a mixed bag in that it was successful for short-term and medium-term debts, whereas the long-term problem was never really solved.[20] It is worthy of note that the Reichsbank had failed to win the day; the new legislation was made comprehensive, so that in effect racial ideology had prevailed over orthodox finance in the Third Reich yet again.

By mid-1935 nearly one-quarter of holdings theoretically eligible had already been rejected, and the total number enrolled was only 684,997 (with an area of fifteen million Ha) by mid-1938. True, this represented rather more than one-half of the land in agricultural use (excluding forestry) in the country. But obviously a surprisingly high proportion of almost one-third of all potential farms had been excluded. This was essentially the fruits of the provisions regarding eligibility in the original law. The decision that in order to qualify a holding had to be large enough to make its owner independent of other revenue entailed that local soil and climatic conditions would be the determining factors, and these varied enormously from one place to another.

Consequently many farms over 7.5 Ha in size were rejected as they simply failed to constitute a viable unit according to the Act's criterion. Apart from bad soil conditions this stemmed from the fact that the holdings were often so fragmented that they simply could not be farmed efficiently. In the former *Realteilung* regions this was particularly true. At Ailringen (Württemberg), for example, of thirty holdings disqualified only one was under 7.5 Ha in size; of the thirty-six rejected in nearby Eberbach, all were over the minimum size, one being 14.5 Ha. For this whole area under the jurisdiction of the *Erbhof* court at Künzelsau 400 holdings were declared unacceptable of which over 80 percent failed to prove economic viability. Similar conditions obtained elsewhere in the Reich; in Bavaria one local court declared flatly that no holding under 10 Ha in its area would even be considered for enrolment.[21]

Many others were excluded on the grounds of being under joint ownership. Again, this was truer of the *Realteilung* districts than of the others: at Künzelsau over fifty of 400 disqualified holdings fell out because of this. So often was it the case that the law had to be repeatedly modified, as in areas such as the Rhineland so many potential *Erbhöfe* would otherwise have been disqualified. As early as October 1933 common ownership was now accepted in respect of married couples farming property jointly. Another amendment followed two months later allowing the registration of holdings partly in such possession and partly owned individually by either the husband or wife. Originally such farms owned jointly went to the man on the wife's death but to the next heir when the husband died first the modification permitted the co-owners to name one another as direct heir, which strengthened the wife's legal position quite considerably, as she could now take over from her husband when she became widowed. The concept of eligibility for commonly-owned properties was extended still further in 1936 when two farms jointly managed were allowed to qualify.[22] These modifications had the effect of considerably increasing the number of *Erbhöfe* in certain regions; the ordinances of 1933 alone added 90,000 properties to the roll. In the Düsseldorf and Cologne districts 40 and 68 percent, respectively, of all registered holdings were only allowable under these amendments.[23]

The lack of efficiency on the part of the farmer seems to have been another ground for non-eligibility: in one parish five holdings out of sixteen acceptable in other respects failed to qualify for this reason. This criterion in itself worked against the concept of a comprehensive law at least to some extent, as by 1 January 1935 2,594 owners had been so disqualified.[24] In many cases the owners themselves objected to enrolment, either for the whole farm or a part of it. Disputes over entry or non-entry on the roll are hard to analyse, as the original sources do not show exactly how many *Bauern* were objecting to entry and how many RNS officials to non-entry, since both types of objection have been added together as one figure. But it seems clear from an examination of the files of individual *Bauern* from various parts of the Reich that no small number attempted to find some way of avoiding enrolment. The number in general was particularly high in those areas formerly enjoying *Realteilung,* such as Cologne and Düsseldorf, and regions in the south-west, e.g. Karlsruhe and Zweibrücken. So marked was this that even the Ministry official responsible for the administration of the law drew attention to it. He estimated that whereas objections in areas formerly with *Anerbenrecht* equalled approximately 10 percent of all enrolled *Erbhöfe* for the same district, in the former

Realteilung regions the proportion was three times as great.[25]

It was partly because of this general tendency that the proportion of apparently eligible farms actually registered varied so widely from one part of the country to another. North and east Germany and Bavaria had a high proportion, especially Schleswig-Holstein, Hanover and Oldenburg in the north-west (where the NSDAP had been well-favoured electorally prior to 1933). Assessment of the EHG's reception and eventual impact must be interpreted in the light of this factor. Perhaps part of the electoral success in these very areas was due to the fact that the local peasants were already accustomed to *Anerbenrecht* so that NSDAP plans for reform of the inheritance laws were less strange to them than to peasants in *Realteilung* districts.

Although public opinion was hard to estimate with accuracy in the Third Reich, some evidence can be offered as to how the new legislation was received. There were genuine protests and the official reactions to them, especially the often defensive tone of NSDAP speeches on the subject, are informative. Local government officials' reports often contained quite frank criticism, since the documents were not intended for publication. In the final analysis how the peasant saw the Act was personal and therefore generalisations are perilous. The EHG offered security in return for restrictions on testamentary freedom, etc. but the vigorous peasant had little need of protection and debt-relief, and therefore received less for what he gave up. As Darré pointed out in a defensive speech on the subject 'It is endlessly said that we created conditions of compulsion with the *Erbhofgesetz*', but he went on to point out that in a nation of sixty millions one could scarcely legislate to suit every individual. The point is well taken: had the government not done something to protect the peasantry it would doubtless have been criticised by the agrarian sector for its omission.

The restriction on *Abfindung* caused most concern, and seems never to have been wholly accepted by the peasantry, as it was so contrary to custom. It was exactly this point which occasioned most comment when the *Bäuerliches Anerbenrecht* was originally promulgated in Prussia. In Württemberg there was a similar reaction.[26] Feeling on the issue had always run high and in one respect had been a grievance prior to the NSDAP accession. In 1929 the police reported in Schleswig-Holstein that limitations placed on *Abfindung* by the recession were a standing complaint among farmers in the province.[27] The NSDAP was treading on rather perilous ground here, in other words this part of the

Act was always unpopular in the Third Reich. In 1937 the president of an *Erbhof* court at Munich wrote to the Ministry of Justice requesting that the courts be allowed to take a more lenient view of the matter, since otherwise the disinherited just left the land for building work or industry. When the LBF for Hesse described local peasant attitude towards the EHG in 1938 as 'dissatisfied' and 'critical' he cited *Abfindung* as the main grounds for discontent.[28]

Against this the party could reply that it had no objection in principle to the disinherited leaving the land anyway, as the peasantry were theoretically the life-source of the whole people and had, therefore, to replenish the declining population in the cities, as well as preserving itself biologically. Some degree of rural migration was acceptable, indeed, actually desirable from the NSDAP standpoint. Moreover, logic was on the side of the party here. Since in the past, many farmers had been run into debt by *Abfindung* obligations, it was a rational decision to forbid them to take on loans in future in this respect. Here again the whole was given priority over the individual.

Abfindung prohibitions must be seen in any case in conjunction with the regulations regarding closed inheritance, since it was their joint effect which was so considerable. The latter provision also came under heavy fire in itself, in particular from Max Sering, the agrarian publicist, and von Rohr. The former wrote a critical pamphlet in 1934, and the point was taken up by von Rohr, who described this clause as the worst part of the Act. He even went so far as to say that it would threaten the whole future of the *Erbhof* as a family concern. Von Rohr believed that the farmer's wife would henceforth keep to herself any land she brought to the marriage as a dowry, rather than amalgamate it with her husband's. This would ensure that something was left to those children not nominated as heir, and would have the effect of keeping the holding smaller than it need be. He also pointed out that indebtedness was normally incurred whilst the children were growing up and subsequently redeemed by their cheap labour as adults. The EHG would militate against this arrangement by driving them away when they did grow up; if they did remain, denied both *Abfindung* and land, they would demand high wages. Again, the point about children being driven away is invalid, as the NSDAP always accepted that some would leave.

Von Rohr's other line of attack was concerned with future credit facilities, based on the logical inference that if the creditor cannot foreclose he is not very likely to risk his capital in the first place. This would have implications for the long-term efficiency of German farming. In February 1934 one farmers' representative in Schleswig-Holstein had already written to Hitler stating that credit had now been cut off from the *Erbhof*: the Regpräsident of Brandenberg confirmed in the following July that borrowing was now much harder than before for the peasants.[29] The official answer to this was that this was a transitory problem only; in the long run increased revenue from higher prices and bigger output would make the *Bauer* independent of outside credit.[30] In defence of the official viewpoint of the matter it has to be accepted that the EHG was based upon the supposition that the agrarian price-index would be kept at a high level; in fact, the lack of outside credit facilities experienced in 1934 made no difference in the long run, as investment actually rose (of which more later).

Lively concern was also caused by the disparagement of women under the EHG, even when due allowance was made for the transitory concession regarding the daughter's claim to inheritance. As late as 1937 one writer admitted that doubts and misunderstandings still existed on the land about such discrimination; these were being fed by the 'evil-minded', who sought to convince the farmer's wife that she had no rights under the EHG's terms. This included the allegation that she forfeited any dowry land which she brought with her.[31] It seems to have been the position of the wife, rather than that of the daughter, that was the issue here. None the less the latter were also affected by *Abfindung* restrictions, as the absence of both land and money influenced their chances of finding a husband. In 1938 complaints in official circles regarding unmarried daughters repeated that it had been overlooked that without land they had no marital prospects; the variable handling of this point by the *Erbhof* courts was said to have caused 'considerable unrest' according to lawyers' reports.[32]

Perhaps the sorest point of all for the peasants was their new status under the Act. Dispossession of the inefficient in effect reduced the *Bauer* to the role of manager on his own holding. One *Erbhof* court judge stated that he knew of cases where parents were being told by their children that they had learned at school that the *Bauer* was not legally owner of the soil any longer. The publication of a guide to citizenship for the young provoked a lively controversy on this subject. In referring to the *Erbhof* it stated: 'It is a form of life which the state gives to a *Bauer* and his kin. The *Bauer* has to carry out the work allotted to him by the state.' This description, smacking of feudal tenure rather than of ownership rights, aroused the ire of Haidn, the leader of Department I of the RNS, who wrote to the editor to protest. He did admit that the passage had a kernel of truth, which is interesting, but suggested that the peasant concept of ownership was not 'liberalistic', as the *Bauer* did not regard his farm solely as an object of material value. These remarks were obviously in line with similar statements made in 1933 by Darré about the supposed difference between *Bauer* and *Landwirt* already discussed here. There was always, to put it no stronger, a certain naïvety in the RNS's belief about the peasantry and its alleged lack of materialism. However, the strength of Haidn's reaction to the offending passage, of which he demanded suitable revision, is none the less revealing.[33] His corporation was clearly sensitive to criticism on the matter.

Furthermore, he conceded in his letter that there was an inner resistance in all parts of the Reich to acceptance of the legislation, an important admission for 1938, by which time one would have expected resignation to a fait accompli, if nothing more positive. Contemporary correspondence confirms that peasants were in many cases reluctant to register their holdings under the Act, except where they were heavily indebted, and therefore had a good deal to gain by enrolment. Indeed, even official sources spoke of 'opposition' or 'bitter resistance' to characterise the peasants' attitude.[34] Surely in this connection it is significant that the Prussian *Erbhof* court at Celle had six panels of judges of which two were solely for objections to enrolment, which clearly indicates a considerable body of litigation on that score. Again, however, warnings must be sounded against over-hasty generalisations, as in one part of the country at least it was said that effects of the law had not been so bad as originally feared, due to the efforts of the local *Bauernführer.*[35]

Even with this reservation it is undeniable that a groundswell of protest was audible in the first months after the announcement of the law. Darré found himself first under pressure on the subject from von Papen, who told him that great unrest had been caused by the EHG. He promptly replied asking for details and saying that reports of this nature when queried turned out usually to emanate from either estate-owners or lawyers.[36] This defence he repeated in a speech to peasants in Bavaria later in the same month. Set against evidence of peasant reluctance to enrol, this line does not sound convincing. His biggest headache however, was the Sering pamphlet, which even became an object of press discussion in various neighbouring countries. The result was the confiscation of the pamphlet by the Minister of the Interior, for which Darré subsequently thanked him. He was later to claim that a Gauleiter's meeting in December had shown that the public was not really concerned over the EHG.[37] If that had really been true, it is difficult to imagine exactly why Darré should have found it necessary to ban all discussion of the legislation at this time, even within the ranks of the RNS itself.[38] His whole attitude was defensive in the extreme and his attack on Sering's views as 'economically orientated' merely echo the usual arguments about the non-materialistic *Bauer.*

It was above all the element of compulsion which turned people off the legislation as well as restrictions on *Abfindung.* In terms of status the peasant now became a farm manager under a quasi-feudalistic contract with the community, the terms of which allowed him to be removed when necessary from having any further say in the management of the property. Provisions on *Abfindung* were tantamount to twisting the knife in the wound. Against this, there is little doubt that many agrarian circles welcomed the legislation in so far as it produced conformity of practice for the whole country. Six months before publication of the EHG a Landowners Association delegation visited the Chancellery and stated that a single law for Germany was urgently required.[39] It is interesting to note that a Bill to introduce compulsory entailment for the peasantry was defeated in the Reichstag during the first war, and that obligatory legislation in Hanover had to be abandoned in 1866 in favour of a voluntary system. The reason for the change was *Abfindung* itself; Hanoverian peasants began to give lump sums from their capital to meet the obligation. This had tended to hinder economic development in agriculture in the region.[40] In view of Kerrl's Lower Saxon antecedents this may have influenced him to hold out during the April 1933 debate for vocational training only at the heir's expense. Equally, the determination of the peasantry to give *Abfindung* when they possibly could testifies to the strength of their feeling on the matter, and consequently makes resistance to the new measure even more comprehensible.

In the final analysis, was the new legislation really justified? There were several obstacles which the peasantry could reasonably have been said as a class to be currently facing. In the long term there was the question of how it was likely to fare in the face of economic developments in the modern world, which tend more and more towards urbanisation. However, curious as it may seem, this was an overrated danger for Germany. The number of smallholdings between 2 and 5 Ha in size had been declining in numbers, indeed between 1925 and the NSDAP's accession more than 50,000 in this category disappeared.[41] A similar trend was observable among the estates of over 100 Ha in size. But in the same eight year period the number of farms between 5 and 100 Ha increased by some 65,000. In other words, those classifiable as *Erbhöfe* were already thriving by comparison with those which lay outside the limits of the new legislation. This scarcely suggests that the peasantry as such was in real danger of extinction. The impression is heightened by the fact that whereas over the time-span 1895-1925 the holdings between 20 and 100 Ha had diminished quite sharply in total, the decline had been arrested and even reversed in the following eight years.[42]

For the problem of fragmentation in *Realteilung* districts, however, the new law did achieve its purpose. In those areas the number of *Erbhöfe*, even though healthy enough to begin with, would have been made larger still in the long run. Further divisions of the soil would now be prevented.

A major point to be considered in weighing-up the Act was indebtedness, of which the NSDAP had always made great play, as it rose so sharply under the Republic. Total farm debts in 1932 were far in excess of current annual output in agriculture.[43] But although the holding at least was now safe under the EHG it should not be overlooked that the *Bauer* as an individual was less secure. As a contemporary commentator realised, one form of discipline had simply been replaced by another.[44] Moreover, indebtedness figures for the whole sector can be misleading, as they may conceal obligations within the agricultural community itself. One peasant in a parish near Wuppertal had registered debts of RM 17,630, but four-fifths were owed to a neighbouring peasant. Another in the same village had no creditors outside the agrarian community. Of fifty-six peasants in Hemmerde (Westphalia) only twelve even claimed debt-relief.[45] By taking the total indebtedness of the agrarian sector as a propaganda point the NSDAP was really misrepresenting the situation.

In sum, despite the world economic crisis, which undoubtedly produced hardship in many individual cases, it could hardly be said that the average peasant holding was in real danger of annihilation in 1933. NSDAP talk about the evils of speculation rested perhaps rather more on preconceived notions about the risks of liberalism and frankly romantic ideas about the *Bauer's* lack of materialistic spirit than on a really vigorous economic analysis.

Similar considerations might well be said to apply to the academic point relating to the supposed revival of old Germanic tradition with the EHG. The Third Reich produced a flood of enthusiasm about the German peasant's historical attachment to his land, and to the old laws which had governed it in the past. Opinions seemed to differ about these latter, however. Sering maintained that *Realteilung* was as Germanic as *Anerbenrecht*: another authority, Brentano, held that the latter form was really Norman in origin.[46] It would appear that uniformity of practice among the *Bauer's* forefathers was largely a myth. There is nothing inherently wrong in myth-making. Historians and social scientists have drawn attention to how fruitful such a version of history can be to a people in enabling it to acquire a sense of identity and traditions to live up to. But there is no real reason to suppose that this particular myth was beneficial or necessary to the German people in 1933 anyway. If the party wished to preserve the peasantry by legislative means then it needed only to say that it had practical grounds for doing so in the light of current events. The business about old Germanic tradition was a harmless piece of romanticism added by the enthusiasts in the RNS.

9

AN EXAMINATION OF ACTUAL LITIGATION UNDER THE ERBHOF SYSTEM

Since a new law had been created for the Bauer it was decided to set up a new chain of legal channels to administer it.[1] The first was the local court (*Anerbengericht*); by 1935 there were no fewer than 1,626 of these in the Reich. The presiding judge was a professional jurist, named for a calendar year, and assisted by two local *Bauern.* These latter were designated by the KBF and called up on a rota system, so that the professional judge did not always have the same two. The authorities were very proud of this 'trial by peers' element in the system and claimed it as a new concept for Germany.[2]

The second link in the chain was the Regional Court *(Landeserbhofgericht).* Prussia was so large that in addition it possessed a third level at Celle, under which were fifteen Regional Courts. There were three professional judges on each panel there and at Celle, assisted again by two *Bauern,* named by the LBF on a rota basis. The highest court was the *Reichserbhofgericht* which acted as the ultimate court of appeal for the whole country. The president was the Minister of Agriculture, who could nominate a professional judge to represent him. The whole panel consisted of three professional jurists and two *Bauern,* the latter installed by the Minister of Agriculture for a period of three years; the RBF was to propose suitable persons for the job to the Minister (who was of course the same person, Darré).

All cases were heard in camera but the *Bauer* appearing was allowed to bring a helper who could be a lawyer. The KBF was always the representative for the authorities in the local court, except in cases dealing with *Abfindung* which was held to be a private family affair so he was not then required. Incidentally, all cases on this issue were settled at *Anerbengericht* level, with no appeal to a higher court.[3] This provision was probably a tacit admission that the higher instances would have become submerged had appeals concerning it been permissible.

At the Regional Court, the LBF was always heard. If the *Bauer* lost in the *Anerbengericht* he had a month to protest formally and demand a hearing in a higher court, the instrument being a written plea signed by a lawyer. Similarly of course the RNS could protest against an unfavourable decision on their side within the same period. Which ever party failed to win at Regional Court level followed a similar procedure to go to the *Reichserbhofgericht* if he desired.

In theory all the professional judges appointed were quite impartial, and not necessarily even party members. For example, at Celle one assistant judge was said to have been a Freemason.[4] But whether National Socialists or not, all court personnel were obliged to attend courses of instruction run by the NSDAP, which were both legal and ideological in content. One session in May 1939 included such themes as property concepts in the Civil Code and in the EHG, the land inheritance question, race, foreign policy and Point 19 of the original NSDAP programme, organised by the Gau leadership. So frequent did this indoctrination become that the Ministry of Justice had to complain to the party authorities about it, as judges were being taken from their work rather too often.[5] In sum, it would appear that the policy adopted by the NSDAP was not so much the appointment of party members only as judges as the conversion of existing jurists to its own viewpoint.

Interventions by the political leadership were apparently never made in the actual courts. When a difference of opinion arose with the RNS it seems to have been the latter which won as far as the actual course of legal administration was concerned. When in Bad Segeberg (Schleswig-Holstein) a *Bauer* who failed to pay either his debts or his taxes was removed from the farm, the KBF later proposed to reinstate him, to which the local party leadership objected, as the man was politically suspect. The KBF promptly rejoined that denial of management rights could not be used as a political weapon, and the party gave way in view of the labour shortage.[6] Here was a clear instance of the RNS refusing to allow politics to infringe on the domain of professional administration. Equally, when an *Anerbengericht* declared a peasant unfit to farm, contrary to local party views, the official concerned left the last word in the matter to the court, with the words 'I restrict myself to producing the Kreisleiter's opinion; it goes without saying that I neither can nor wish to intervene in the unfinished legal procedure'.[7] It would seem that the party tried to indoctrinate the judges, but then left them free to decide afterwards.

All professional jurists were appointed in any case by the Minister of Justice, not by Darré: the former, Dr Gürtner, was not a member of the NSDAP having been in previous cabinets as well as in Hitler's. He certainly believed in principle in the impartiality of the law. The ill-treatment of prisoners at Hohnstein in Saxony, and the trial of the SA camp guards involved, led to his letter to Hess demanding justice since otherwise 'independence of the judiciary, which is the basis of an orderly administration, [will] disappear'.[8] There is no reason to suppose that this principle was not adhered to in respect of the EHG litigation.

However, nothing in the regulations prevented *Bauernführer* themselves from being assistant judges, and they were usually party members. The LBF could quite easily have named a KBF to sit in judgement at Celle but in order to ensure some degree of impartiality it was stipulated that no KBF could sit in judgement in a case which he himself had initiated.[9] At the *Reichserbhofgericht* level one *Bauer* nominated to serve a three-year term by Darré turned out to be deputy LBF in Hanover: not only was he therefore an RNS official, but it appeared that he held high office in the SS and had a long party record.[10] It seems very unlikely that anyone politically suspect would have been called up on the *Bauer* rota at any level, so that to a certain extent actual party intervention in the court procedure was unnecessary.

Despite the apparent tendency towards partiality, however, genuine efforts were made to arrive at a fair judgement which accounts presumably for the appeal courts in the first place, since their existence made the course of justice very long-drawn out indeed. An analysis of judgements given by the *Reichserbhofgericht* in respect of Bavarian *Erbhöfe* from February to June 1937 shows that the average delay since the original local judgement was 23 months.[11] The inordinately lengthy process was in general determined by three separate factors.

Of these, the most important was legal overwork. The Prussian court at Celle dealt with 8,293 cases, only two-thirds of which had been settled, in its first ten months. New appeals were arriving at a rate of

200 per week.[12] This was far more than the court could comfortably deal with; the Vice-President of the *Reichserbhofgericht* estimated that no panel of judges could properly assess more than 750 cases annually.[13] As there were only six panels at Celle they were clearly over-burdened.

The second obstacle in the way of quick decisions was the nature of the work itself. By definition, the technicalities of farming were bound to be considered in cases of bad management, or in those instances where enrolment was concerned, since objecting peasants always tried to prove that their holding was not sufficiently rich to make them independent of outside revenue. To decide on this involved listening to technical evidence about soil, etc. and possibly visiting the holding, all time-consuming. It was because this kind of spadework had not been done at local level that the judgements in the examples of the Bavarian *Erbhöfe* mentioned above had been so delayed. Eleven of the thirteen cases had had to be sent back to the *Anerbengericht* for further investigation.

Apart from the quantity and type of work concerned, the third element in delay was often deliberate in that the unscrupulous could exploit all possible legal channels to their own advantage. To cite only one example, a farmer in Hindorf (Saxony) had a 75 Ha holding which was heavily indebted, so he unsuccessfully claimed *Osthilfe* in July 1935. His next course was to try for enrolment under the EHG and so prevent foreclosure. When that also fell through, after a lengthy procedure, he went to the ordinary courts. Eventually in December 1938 he came to an arrangement with his creditors, some three and a half years after the beginning of their campaign to get satisfaction.[14] To such peasants the EHG was simply an additional weapon to be invoked against fulfilling their obligations.

In effect, two kinds of litigation took place under the new law; in the early years there was a veritable flood of processes to decide whether or not the holding in question was an *Erbhof*. This meant that 1934 was a busy year for the judges; by that October the Regional Court at Munich had already dealt with 1,600 cases and was asking for more judges. In 1938 one-third of all litigation so far had involved the question of eligibility for registration.[15] The obvious corollary is that once this matter had been definitively decided for existing farms the number of cases per annum was bound to diminish, so that the amount of work for the courts in the early years is rather misleading. Indeed, the number of holdings enrolled changed only slightly between 1 January 1935 and union with Austria in March 1938.

Once these hurdles had been cleared and the bulk of the holdings registered the courts settled down to a rather different routine. There were then four main causes of litigation: attempts to sell or exchange land, objections to registering newly-acquired pieces of property as part of an existing *Erbhof*, the question of a mortgage on the farm, and finally problems regarding the handing-over to a successor.

One preliminary point is that an ostensible cause may well conceal a real one. This has particular importance when *Abfindung* is considered. It seems very probable that when the *Bauer* tried to sell a portion of his farm he had in his mind the desire to obtain liquid cash for his children. Equally this applied when he tried to get one part of the holding left outside the provisions of the Act, as peasants so often did.

From the samples assessed there are two generalisations which appear fairly safe to make. First, there was a high degree of involvement with the law in all regions, the usual proportion of *Bauern* who had recourse to the courts being somewhere around 50 percent of all those in that neighbourhood: in itself, this illustrates how closely farm management was supervised. Secondly, it appears that the overwhelming majority of cases were settled at *Anerbengericht* level. The impressions gained therefore are of a close surveillance of the holding, and of the apparent acceptance of the local decisions. Knowledge of how long appeals would take may have played a certain part in this. But on the other hand it may equally be testimony to the court's willingness to conform to local customs in interpreting the law, which in itself meant that the authorities were understanding in their attitudes, at least in the early years. A further factor here is *Abfindung* and its frequency as a cause of dispute; in such cases there was no appeal possible anyway.

In principle it was forbidden to burden the holding with a debt in respect of *Abfindung*, but in practice the courts often showed appreciation of the owner's desire to do this, where the farm could bear the proposed mortgage. In the district of Hameln-Bad Pyrmont (Lower Saxony) the court was very lenient in such cases, and almost always accepted reasonable mortgages. When the youngest daughter inherited a property at Brockensen, for example, the father was granted a loan against the farm to give the eldest daughter some money. Similar instances of understanding of an individual's desire to see his family settled financially were forthcoming in respect of land sales especially if arrangements regarding the strip in question had been made prior to the Act. A *Bauer* in Unna (Westphalia) was allowed to sell land for *Abfindung* as he was able to show that he had originally acquired it specifically for his four sons before 1933.[16] As far as former *Realteilung* districts were concerned it was eventually accepted that the new provisions were harsh, as all children were accustomed to receiving some land. Consequently a state subsidy of RM 5 million was provided to cushion the shock. Benefits were limited to RM 5,000 per person.[17]

But court decisions were variable, particularly so where land sales were concerned, as the holding was supposed to remain inviolate. In one instance in Schleswig-Holstein a *Bauer* tried to sell a piece of land to give vocational training to his seven sons which the *Anerbengricht* rather surprisingly sanctioned. The KBF immediately objected on the grounds that a precedent was being created and eventually the case went to the highest court. The ruling stated that: 'If liquid resources are not available to make the other sons independent their training can be paid for gradually through the current surplus of the holding.'[18] What the owner was supposed to do if the current surplus was not big enough was left very much in the air. The corollary of this may well have been that the bigger the farm the less unfavourable these provisions on *Abfindung* and vocational training were.

Although vocational training of the offspring not inheriting the land was a legal obligation, the local *Bauernführer* might still object if the sum allowed was too heavy for the holding to carry. He could take the case to court and if the judges agreed a new contract would have to be drawn up between the heir and the other siblings. The potential burden of this expense could sometimes even serve as an excuse for non-enrolment of the holding. One peasant in Ailringen (Württemberg) successfully objected on the grounds that he had thirteen children, so that any heir would have had to finance the vocational training of the other twelve, which the holding could not stand.[19] The fact that he would not even have been allowed a mortgage for the purpose has to be borne in mind here.

Sales of *Erbhof* land were occasionally allowed where it could be shown that disposal was in the interests of the community. One *Anerbengericht* permitted a *Bauer* to sell a strip to a post-office official who kept pigs as this encouraged small-scale settlement. Equally if it assisted efficient farming owners could sell parcels of land to one another, as in one case where the transaction allowed one holding to be cultivated as a block. Community interest covered sales as building sites or for roads as well which turned out to be highly lucrative. An OBF in rural Lower Saxony offered a plot to a company for a housing estate at a rate of RM 10,000 per Ha. Inflated rates seem not to have been unusual, as another *Bauer* in Schleswig-Holstein got the equivalent of RM 5,000 per Ha as early as 1935 when he sold a paddock for construction purposes.[20] This is an ironic comment on the agrarian policy in the Third Reich in general, because after all the object of legislation such as the EHG was to eliminate land speculation. When high prices were available, however, the *Bauer* appeared to be as keen to profit as anyone else. NSDAP propaganda about his lack of materialism seems curiously naïve beside the actual facts of land sales.

None the less the sale of an entire property was rare although again circumstances were sometimes allowed to alter cases. When one *Bauer* lost his farm buildings through a road scheme he was given permission to sell off all the land in parcels so that he could acquire a complete new *Erbhof*; also the sales would create one or two *Erbhöfe* in the parish by enlarging the farms of the purchasers. Disposal of the entire holding had to be accepted where no heir was available, as in one instance in Westphalia. Twelve *Erbhöfe* were sold in Bavaria over a six month period in 1936, so that although clearly it did not often happen, none the less on occasion the holding ceased to in the possession of the original family.[21]

In other words, whatever the letter of the law the judges usually adopted a pragmatic approach to such transactions, so that the legislation in this respect was never so restrictive as a casual reading of its provisions might imply. The needs of the community and the individual circumstances were frequently given precedence over more legalistic considerations. Sometimes this flexibility of interpretation assisted the family but not always. One peasant obtained permission to sell the whole farm, which he said he had never run himself, despite the objections of the relatives. The court ruled that only a direct heir could complain about such a sale, and that the holding should be allowed to 'bloom again' under a competent owner and contribute to national output.[22] There was no question here of 'Blood and Soil'; Germany's drive to self-sufficiency took priority over family considerations.

Pragmatism also ruled very often in cases of division of a holding. Again in theory this practice was inadmissible but farmers tried to achieve it none the less via legal transactions. In one example a *Bauer* divided the 35 Ha holding among his three sons, which the *Anerbengericht* accepted. Then the KBF got the decision reversed at Celle and the *Bauer's* appeal to the *Reichserbhofgericht* was turned down, but only because the two smaller allotments would have had no buildings and therefore could not qualify as *Erbhöfe*. Apparently, had the father been able to afford such facilities the arrangement would have been approved, so that a considerable degree of latitude in the interpretation of the EHG's provisions against division can be seen in actual practice. In Oldenburg an owner was permitted to bequeath a farm of 35 Ha in the form of two roughly equal holdings to his sons, provided that he enlarged the smaller by additional purchases and provided the necessary buildings.[23]

Equally flexible were the decisions made in respect of small pieces of land allotted direct. In Bavaria the court allowed a *Bauer* to give nearly 2 Ha to his daughter on the grounds that she had been promised it in 1920. However, decisions were extremely variable, depending on the court and the individual circumstances. In order to circumvent the provisions about indivisibility the owners often tried to register part of their farm only, with some property left freely at their disposition; this did not always work, although often it took a long legal battle to decide. A *Bauer* in the Rhineland tried to leave a piece of his land unregistered, so that he could pass it on to his daughter, which the two lower instances both accepted, despite the KBF's objections. The LBF promptly went to the *Reichserbhofgericht*. Here he obtained satisfaction; the supreme court ruled that if the piece intended for the daughter were instead sold to her husband, his enlarged holding would be viable as an *Erbhof* in itself. In other words, the chance to create a new hereditary farm for Germany overrode personal wishes.

This kind of ideological thinking often governed court decisions, where any possibility of swelling the ranks of the *Erbhöfe* could be seen to exist: on such occasions all personal rights just went by the board, in accordance with the general National Socialist slogan of 'The common good before individual gain'. In one such case in Brunswick a widow was not allowed to farm two holdings as one unit after her husband's death, although they had been in practice since 1916. As one of the two could have become a separate *Erbhof* in its own right, the courts refused her permission to amalgamate them. Sometimes denial of normal property rights went even further, as when a *Bauer* in Nahe

(Schleswig-Holstein) tried to leave his daughter a small house adjoining the farm, but after KBF objections the *Anerbengericht* ordered the house to be included in the holding's registration. At Celle the KBF was upheld.[24] As the property in question had been acquired in 1922 one could well imagine the farmer's feelings on the matter. What makes this particular process so glaring an example of the individual being pushed to one side was the fact that after all the whole case was about one building, not actual farm land. Apparently the KBF's fear that the daughter might have sold this property to the (unexplained) detriment of the holding decided the issue.

One of the most frequent causes of litigation was the contract drawn up when an heir took over. There were various reasons why the KBF might object and ask the court not to accept the document as devised. It might come into dispute if the retiring *Bauer* nominated someone other than the expected heir; although this was allowed under certain circumstances it was up tho the *Bauer* to prove that grounds for the deviation existed. A court ruling in Westphalia showed that the judges were willing to comply with a desire of this nature when there were good reasons: one man was allowed to designate his grandson as his only son was not a capable farmer.[25]

What complicated contracts was the condition which was theoretically the very basis of the EHG, that all *Erbhöfe* should remain wherever possible in the same family line. Litigation often became extremely involved on this issue, as sometimes practical considerations seemed to suggest that this clause should be overridden. One holding was allowed to be transferred as the Celle regional court held that the contract bequeathing it to a married daughter was valid, since her husband was so efficient. The latter was farming both his own land and his wife's as one unit, so that her inheritance would be lost eventually to her own family line. Acceptance by the court of the contract meant that here again economic considerations were given priority.[26]

What caused the greatest amount of work for the courts in respect of contracts was the need to establish local inheritance customs. This meant that inquiries had to be made locally to determine the district's tradition since custom varied sometimes from one parish to another. In Schleswig-Holstein this question caused immense trouble for Celle and took up an inordinate amount of time, as the opinion of the local worthies was taken as the criterion, which involved frequent consultation with them. Just how varied custom could be seen from a report on

Elpersbüttel; of fifty-four *Erbhöfe* there, twenty-three had been bought by their present owners and the remainder inherited. Of the remaining thirty-one, eighteen had been passed on intact to one heir, eleven divided, and two were inherited under terms of joint ownership. Judging from the chairman of an *Anerbengericht* reporting to Celle, this was not untypical of that area of Schleswig-Holstein. On the other hand, in Lauenberg and Stormarm the overwhelming majority of parishes had primogeniture.[27] These variations appear to throw some doubt on the EHG's claim to have revived old Germanic custom in closed inheritance, as clearly the real situation prior to 1933 had been far more complicated.

In Nassau yet another custom was in vogue whereby when the peasant died intestate his wife enjoyed the use of the property during her lifetime. In those cases where arrangements of this nature had been made prior to the EHG they were allowed to stand. When the wife died then the holding automatically came under normal *Erbhof* law.[28] From the foregoing it can easily be estimated how much time and effort was devoted to establishing customs: as a result in 1939 the REM and the Ministry of Justice published a joint decree to the effect that they were authorised to establish what the tradition had been in each district when the EHG had come in and publish it: this then became binding for that area. In this way it was obviously hoped to remove some of the burden on the courts.

There are two main points of interest in this whole matter; first the judges clearly did take pains to determine local practice, which meant that the application of the inheritance law varied accordingly; it would therefore be untrue to say that every *Bauer* now had to conform to standard requirements. In this respect the EHG was less arbitrary than might at first be suspected.

Secondly, there is the question of *Bauernführer* influence behind the scenes; as already pointed out they were quick to protest against contracts they did not care for, and in 1935 it was alleged that they were actually going even further and persuading the peasants to agree to conditions not even called for in the EHG. This brought an angry denial from an RNS official, pointing out that many contracts had to be turned down by the courts precisely because they did not go far enough to satisfy the EHG's conditions;[29] this is a valid point but it is unlikely that a lawyer would have made such a charge unless he had some information or experience to substantiate it. The power granted to them was considerable, and it is difficult not to believe that they exercised great influence in the court hearings. Here as in other respects the local *Bauernführer* was the important rural figure in the Third Reich.

In no respect was testamentary freedom so limited as in the provisions regarding female inheritance. The open discrimination against women in the original Act was actually reinforced in 1936 by an amendment allowing the *Bauer* to declare by deed that his property was closed in the male line in perpetuity.[30] But the courts were quite flexible in practice on the subject of female succession; in Dresden a *Bauer* who was childless adopted his brother's daughter and named her as heiress; his three brothers fought this to the *Reichserbhofgericht* and lost, the judges upholding the girl's claim upon the length of her domicile with her uncle. In Ingelfingen (Württemberg) a daughter was given preference over a son at the father's request. It seems clear that the *Bauern* had always regarded daughters as capable of inheritance, even in cases of small pieces of land, which may have been intended as dowries. A peasant in Bavaria gave his daughter a strip of under 2 Ha in extent, leaving only 7.5 Ha for the rest of the *Erbhof*, which the *Reichserbhofgericht* accepted, as the arrangements had been made in 1920 and the rest of the farm still constituted an *Erbhof*.[31] This again was a very reasonable decision, made against KBF objections, and shows court willingness to respect existing arrangements during the transitional period after the EHG's introduction. The depth of feeling on the whole subject can be gauged from the decree of April 1939, which extended the initial concession regarding the promotion of the daughter's claims over those of the testator's father or brother to the second succession as well as the first. Clearly the NSDAP desire to discriminate against women was being mitigated in actual practice, and probably by peasant opinion.

Some litigation inevitably took place over the possible removal of the owner from farm management. The process had two degrees called 'simple' and 'strict'. Under the first the *Anerbengericht* could designate an heir to whom the property should go as a kind of trustee until the present owner's death; a wife could be so entrusted in default of a normal heir. Under the stricter version the holding went over immediately to a new owner suggested by the RBF and the *Anerbengericht* was bound by his decision. The designated person was not bound to purchase, in which case the farm went up for sale.

Although efficiency and honour were the two criteria applied, in practice it was often hard to separate one from another. One *Bauer* in Schleswig-Holstein who drank heavily, ill-treated his wife and ran his farm so badly that she complained to the KBF was replaced by a trustee for ten years. On other occasions a lack of honour alone was sufficient to deprive the *Bauer* both of his title and of farm management. One contemporary writer on the subject produced evidence from various parts of Germany on conviction in the civil courts of such offences as swindling, theft, attempted seduction, rape, etc. and concluded that 'a prison sentence always has as a consequence the loss of honour'. Indeed in one case suspected arson was sufficient; a *Bauer* in Saxony tried and acquitted of this in the ordinary legal way was still later shorn of his title. The non-payment of outstanding debts was also taken as grounds for dismissal, as in one instance where a *Bauer* in Leizen (Schleswig-Holstein) did not settle up, and the wife was installed as trustee for five years as the designated heir was only seventeen years old.[32]

The possibility of signing a contract with the heir to provide new management was sometimes used by the RNS as a weapon to get rid of the incompetent; the deputy LBF in Lower Saxony told one 88-year-old *Bauer* that if he did not sign an agreement in four weeks he would apply for his removal. In another instance a local savings bank successfully applied for a *Bauer* to whom it was a creditor to be removed formally from management of the farm, but after he signed a contract with the heir the process was cancelled.[33] No doubt the fear of being in effect publicly branded as incompetent was the main factor here in getting the recalcitrant owner to make way gracefully.

The Four Year Plan did produce increased pressure on the *Bauern* for more output but no really large-scale drive against incompetence; it is true that some bad managers were evicted by Celle in 1937 but the proportion was small, bearing in mind that there were nearly 400,000 *Erbhöfe* in Prussia. In one typical case the dispossessed, when charged by the Bürgermeister with having failed to produce more, simply replied 'The harvest is big enough for me': his lack of responsibility led to his dismissal. Erich Koch went on record as saying in April 1937 that in every village in East Prussia there were one or two really bad *Erbhöfe*, which not only did not contribute to increasing output but were counter-productive through being such bad examples: the owners could not be removed, as there was no one with the capital to take them on.[34] This point shows clearly that whatever the legislation laid down in principle, the practice was often virtually bound to be different. Therefore a really widespread drive against farming inefficiency was scarcely possible, which acted as yet another brake on literal implementation of the EHG.

In any case the NSDAP could reasonably reply to critics that it had not exactly made any secret of its principles on land ownership. They had been put quite explicitly in the party programme in March 1930, 'The right of property is bound together with the duty of utilising the soil for the good of the whole people'. Of course this was a general principle rather than any sort of really detailed proposal, but it is hard not to feel that any peasants who chose the NSDAP had been fairly warned in advance that this was a monistic movement. Equally Willikens had written that the party did not demand that the individual should surrender his property, but merely the right to do with it what he wished, irrespective of the community's needs. These sort of sentiments dispose of the idea that the NSDAP was just another Right-Wing party, but equally they suggest that those peasants who did vote for Hitler had been warned in advance what it might entail in some respect at least.

Obviously another aspect of this whole question of status is the almost stiflingly close supervision to which the *Bauer* was subjected, even when left in charge of the holding. Credit facilities are another illustration of this tendency. When a *Bauer* with just under 20 Ha tried to obtain a mortgage to buy some woodland, the KBF objected as he already had some debts and the acquisition of the new property was not essential to good farming: the *Anerbengericht* refused the applicant permission to purchase. On the other hand when a peasant needed money to carry out repairs to his dwelling-house this was accepted as being necessary, and permission for a mortgage was granted. The KBFs were apparently mainly concerned with the distinction between long- and short-term credit facilities, since the official in Tremsbüttel (Schleswig-Holstein) complained when the *Anerbengericht* sanctioned an instance of the latter; at Celle the decision was reversed, with the grounding that an *Erbhof* should pay short-term capital requirements from its own resources and if that was impossible, the owner should take up personal credit.[35] With this degree of surveillance it is not surprising that by 1936 courts had sanctioned only 13,091 farm mortgages for the whole country.[36] As there were nearly 700,000 *Erbhöfe* permission to burden the holding was obviously not lightly granted.

Apart from new debts the existing commitments were important, especially in determining whether a property could be registered under the Act or not. There was for the heavily-indebted an obvious advantage in getting enrolled as their creditors could not then foreclose. There were some quite shocking instances of payments delayed in this fashion; one peasant in Ailringen, who had managed to accumulate debts of RM 25,000 on a holding of less than 7.5 Ha, held out until June 1936 by constantly appealing for registration. This was on a farm on which foreclosure had been ordered in November 1932. In a similar case at Sommerach (Bavaria) a would-be *Bauer* had debts equalling six times the farm's value, and took advantage of all possible appeal courts. In February 1937 the *Reichserbhofgericht* referred the matter back to the *Anerbengericht* on the grounds that the KBF had not given evidence at the original hearing.[37] These sort of legal delays merely stoked up the fires of exasperation among farm creditors, who found literally to their cost that for the wily peasant proprietor the new Act was a godsend. Consequently a good deal of litigation stemmed from creditors opposing farm registration so that they could foreclose. It is easy to sympathise with the feelings of the exasperated lawyer who wrote to the EHG authorities pointing out that permission to foreclose on an indebted farm was being sought, on the mortgage payments of which his client was trying to exist, although she had received nothing for two years. The solicitor added that that day he had received the usual advice, that no decision could be expected in the immediate future, i.e. as to whether the farm was an *Erbhof* of not.[38]

What was undoubtedly galling to persons in similar circumstances was the lethargy of the *Bauernführer* themselves in following up the matter, possibly because of innate sympathy for the peasants. A Chancellery minute referred to one of the 'repeated cases' where the KBF was not energetic enough in such an issue.[39] Indeed, it was hard to be a creditor in Germany in the inter-war years, as even prior to 1933 the Osthilfe provisions against foreclosure (see page 53) made their life difficult in regard to rural debtors.

Any summary of the legislation has to emphasise that the first point to be made is the variability of its effects according to individual circumstances. Essentially the main item on the credit side was security of tenure, but this consideration would clearly weigh more heavily the less efficient the *Bauer* was. This would apply to debt-relief for similar reasons. It might be argued against this that the less competent (or honourable) owner was more likely to be removed from management; this is true, but the scale of this operation was really very slight. It was always a spur to the lethargic but not one that was very often used in reality. For the less vigorous the provisions of the EHG against foreclosure, etc. were a welcome umbrella under which he could shelter from financial storms, even if his personal capabilities were mediocre. Against this protection he exchanged a good deal of his basic property rights, but on the whole there were doubtless many peasants who found the bargain good enough after the time of the world economic crisis.

Almost certainly the restrictions were most onerous to the more efficient *Bauer*, especially as they were so wide-ranging. He was no longer able to dispose of the holding as he saw fit; as well as needing official permission to sell the whole or part of the holding, or to take up a mortgage, he was limited in respect of the money given to those of his offspring not inheriting the land. Needless to say he even had to go to court if he wished to lease out the farm, provided that this was for three years or longer. If a new piece of land was acquired by an *Erbhof* owner it had to be enrolled as part of the holding, although both he and the KBF had a month in which to lodge an objection. But unless the peasant was prepared to go to law with adequate proof the fresh land became automatically subject to all the restrictions placed on the rest of the farm. These are all irksome restraints to an enterprising man.

Psychologically he had also to accept the new status under which he was rather a manager than an owner in the full sense of the term. Against this he had, it is true, the honorific *'Bauer'* as a kind of compensation for real loss in standing. Moreover, it was the peasantry as such who tended to be the object of enthusiasm in the new state, rather than food-producers in general. To a community suffering perhaps from a certain inferiority complex prior to 1933 the fuss may well have been welcome and has to be weighed in the balance against loss of standing in other respects.

It was not lost on contemporary observers that a comprehensive measure was bound to include the inefficient. As one lawyer pointed out, the tendency to enrol as many farms as possible had shown itself to be false.[40] This is interesting, as after all this issue had been raised by the original begetters of the Act at their April meetings. Perhaps long-standing NSDAP thinking about the need to preserve the peasantry as such prevented the matter from being settled in any other way. As early as 1929 Darré had written that 'If one does not protect the peasantry by special measures a state founded on the idea of profit alone . . . leads sooner or later to the destruction of the *Bauer.*[41] In other words, economic criteria were not to have priority. Not that they were not grounds important also for maintaining a healthy farming community. As Hitler put it in 1933, 'Without the counterweight of the German peasantry the Communist madness would have long ago overrun and annihilated the German economy.' There were equally military reasons behind the EHG, as the agrarian sector was held to be the backbone of national defence. A whole complex web of thought lay behind the promotion of the *Bauer,* the general conclusion of which was that the whole class had to be maintained. This was fully in accordance with nineteenth-century Romanticism and its eulogies of the peasantry as a solid conservative block against modern developments.

Finally the courts' interpretation of the legislation has to be taken into account when assessing the degree of hardship involved. From the evidence of individual cases it does appear that the judges were impartial and that they made every effort to conform their decisions to locally-accepted custom. One historian has even gone so far as to maintain that through the 'understanding judgements' of the courts the EHG was never really carried out in actual practice.[42] This is perhaps a little too strong as a summing-up but certainly an impression of flexibility and a readiness to accept arrangements already made prior to the Act cannot be denied: this particularly applied to land sales of which 54,591 had been sanctioned by the courts in 1936.[43] The facts of economic life were frequently allowed to prevail over theory. Of course, decisions were inevitably aimed at implementing the EHG in principle, within a context of political indoctrination of the judges as described here. No evidence of any direct party intervention has been found, however. These observations have to be tempered by the fact that the immediate transitional period of the legislation has been described here, when concessions were made explicitly for that reason. How the law would have developed in the course of time is impossible to determine now, but it seems rather unlikely that it would have got milder. Equally, it would certainly not have promoted a more efficient agrarian sector in the long run, perhaps the only thing that could have led to its modification. In effect, this is tantamount to the proposition that economic considerations would have had to be given more weight, despite 'Blood and Soil' theorising.

10

THE NSDAP's SETTLEMENT PROGRAMME

Its Principles, Organisation and Achievements

The third pillar of the National Socialist structure to aid the peasantry was a settlement drive in Germany's relatively underpopulated eastern areas. This again was scarcely an original idea, as virtually all political parties had advocated similar schemes. A desire to repopulate the east was born in the late nineteenth century as the solution to the long-standing problem of migration to other parts of Germany. The Stein-Hardenberg land reforms initiated at the Napoleonic period had deprived the one-time serfs in Prussia of about 1.6 million Ha.

After about 1850 the growing industrialisation in the west acted as a magnet to the eastern peasant or landworker. Between 1841 and 1910 five eastern provinces lost a total of 2.75 million people at a time of overall demographic growth for the country. To replace them the estate-owners took on Polish immigrants, a movement which assumed such a degree that Bismarck forbade it in 1886. The ban did not appear very effective, possibly due to Bismarck's resignation, and by 1914 one-seventh of all agricultural labourers in Germany were Polish.[1]

This development caused dismay to the estate-owners themselves, as German workers and peasants turned into an industrial proletariat in the west liable to fall victim to Marxism, whilst the eastern areas filled up with foreigners. As a consequence the Prussian (internal) Colonisation Commission was established in 1886. The object was to acquire large estates in districts with a high proportion of Poles; this land was to be divided up into plots for Germans. Unfortunately the Poles bid for the same estates and pushed the price up, so that the scheme was on a limited scale, but it was a beginning.

In order to maintain morale in the trenches settlement propaganda was intensified in the war; a Reichstag committee was set up in 1915 to superintend the whole matter. In June 1918 a draft bill was published as advocated by Hindenburg as a means of reviving the Army's 'flagging strength'. Early in the following year he issued an enthusiastic statement on postwar settlement plans including the phrase 'hundreds of thousands of places'. An additional factor from 1919 onwards was the Treaty of Versailles which deprived the country of considerable areas in the east. Unfortunately these comprised some two-thirds of all new holdings previously created by settlement prior to 1914.[2] Germany virtually had to start afresh.

This may well have acted as a spur to the Settlement Act of 1919, complemented by a Homestead Act in the following year. It was clear from the new legislation that the main source of land would be the large East Elbian estates, in the possession of the Junkers. Many of these were, however, entailed, hence Article 153 of the Weimar Constitution with its laconic sentence: 'The entailments (*Fideikommisse*) are to be dissolved.' To ensure a ready supply the 1919 Act decreed that in all districts where estates over 100 Ha in size formed 15 percent or more of all agricultural land the owners of these were to surrender enough to reduce their share to 10 percent. Breaking up the estates in favour of peasant farming was by no means an original concept, as Max Weber among others had suggested much the same thing in the 1890s; non-viable estates would make way for a 'numerous and healthy peasantry' to protect the eastern frontiers.[3] Even the settlement programme of Weimar partially stemmed from a desire to 'damn-up the onflowing wave of Slavs'.[4]

Despite such urgent political necessity, however, the expectations of the new Act exceeded results. Theoretically there were 1.5 million Ha available from the estates due under the dissolution of entailments. In practice by 1928 only a little over a quarter of this land had come on the market. The Junkers were using their connections well to impede the break-up of the estates; further decrees were passed by the Prussian government to accelerate the process but to little avail. To the intransigence of the estate-owners was added the friction between administration in Prussia and the national government which further delayed the implementation of settlement projects.[5] Although there is little doubt that the coalition cabinet in Prussia was hostile to the Junkers and would dearly have loved to destroy their economic power-base, a combination of unfavourable circumstances prevented them. The abuses of *Osthilfe* merely exacerbated the friction between the Prussian government and the big landowners.

Coupled with these difficulties was the fluctuation in demand, due to the general economic state of the country. At times of distress, such as in the early post-war years or after the onset of the world economic crisis, demand for places tended to increase as compared to the relatively low interest shown in the middle twenties. As a result the Republic was able to install by 1933 only 57,457 new proprietors on a total area of just over 600,000 Ha.[6]

However, by 1932 public interest appeared to be rising again, due to high urban unemployment and insecurity. A life on the land seemed a better prospect than the apparent lack of any sort of economic future in the cities. Moreover, many felt that if peasants' sons or landworkers were granted a place of their own in the east, then at least rural migration and an increase in the number of city unemployed could be prevented. Consequently even the trade unions sent in a petition calling for settlement as a weapon against rural depopulation.[7] Max Sering himself (the principal framer of the 1919 Settlement Act) estimated that because the production rationalisation half of the unemployed would never be able to find jobs in industry again. Later he was to call for a settlement programme embracing a million people as a remedy.[8]

Here were further grounds for internal colonisation as well as the perennial problem of the Pole at the gate which still retained its old force. By April 1932 Polish land purchases were causing so much alarm that a Prussian governmental conference was called, with representatives from the national government, to discuss the entire issue. In Upper Silesia the problem was especially acute because of the large Polish minority. The fear here was that immigration in conjunction with continuing rural migration by the Germans in the area would threaten eventually even the existing frontiers. Consequently not merely the nationalistic parties but virtually all political movements in Germany formulated demands for eastern settlement; for example, in July 1930 the *Deutsche Staatspartei* (the successor to the DDP) led by Theodor Heuss and Dietrich called for a large-scale project against 'the threat to German land and culture'.[9]

The NSDAP itself had always displayed some interest in the question. In 1924 its group in the Reichstag had put forward a resolution demanding intensified settlement.[10] The March 1930 programme took it a stage further by linking such schemes with inheritance law reform, so that disinherited farmers' sons and aspiring landworkers would find a new holding in the east. A special department called *Ostland* was set up in the Agrarian Office to deal with forward planning. By 1 June 1933 guide-lines for the new policy were given out to all relevant authorities.[11]

The preamble consisted of the usual statement about the need to preserve the peasantry. Only the best racial elements were to be participants in internal colonisation, so that future generations would be strong enough to till the soil. At least one-quarter of the places should go to aspiring peasants from the south and west, so that the previous trend in migration could be at least partially reversed. None the less, other frontier regions should not be neglected, specifically in the north-west and the Bavarian Ostmark. But the main objective clearly was securing the eastern frontier, or as a contemporary put it, build 'a living wall against the Slavs pressing forward'.[12] New villages were to be formed on the basis of farms of varying sizes, although normally the peasant holding was to be preferred. Agrarian re-settlement was to be only part, as Darré envisaged a gradual re-orientation of the whole economy, based on eastern re-population.

The necessary administrative measures were speedily introduced: on 14 July Darré presented a Bill which gave sole authority to the Reich.[13] The NSDAP hoped to obviate the friction which had existed between Prussia and the central government under the Republic. The existing private settlement companies, the instrument for actually placing the settlers, were to be subjected to critical scrutiny: in fact, of the more than one hundred operating in January 1933 over sixty were eliminated as not efficient and the rest came under the Reich (REM).[14] Unity was created in this way with the state as the executive authority. On 14 July the Minister of Labour introduced a bill giving both the regional governments and himself the power to name whole regions specifically as settlement areas: the basic idea behind this move was to promote uniform land-procurement in these districts.

A lively debate took place within the ranks of the NSDAP as to what size of holding should be taken as a criterion. The *Ostland* section had originally suggested the normal Weimar one of 15 Ha, but in 1934 Darré was still seeking advice on the subject from Willikens. The latter was by now ultimately responsible for the whole rural settlement programme. In this section of the REM the prevailing opinion appeared to be that the creation of a 'healthy peasantry' should take precedence over 'striving after record figures'; the new holdings should certainly be large enough to be viable and to ensure that there was no perpetual land-hunger in the owner's family.[15] It is important to stress this as otherwise it may appear that because the NSDAP ultimately did not create as many new places as Weimar its projects failed. In fact, it never did chase after mere numbers but preferred to concentrate on a tolerable size for each holding. Indeed, one justifiable criticism of the Weimar programme was exactly that it had produced far too many tiny farms. Not until 1931 were even half its new holdings as large as 10 Ha, in itself hardly large enough for the east, where even a pair of horses required 2-2.5 Ha for feeding. The non-viability of many such farms was a constant embarrassment to the new régime, which had to grant repeated aid to settlers given places by previous governments. These facts have to be borne in mind when assessing the worth of the NSDAP's plans.

However, even if a small number of new places of 15 Ha each were to be brought about a good deal of land would be needed, a problem which had exercised previous governments. In his guide-lines Darré had envisaged the improvement of waste land and the winning of new land from the sea. These sources of supply, however, were frankly inadequate; probably Darré was really aiming at the large estates. Previous efforts to break them up had possibly played some role in unseating Brüning: but that the estates were seen as providers at the time of Darré's accession can be easily deduced from the calculations of the governmental Settlement Planning Office. In December 1933 it reported that if 90,000 projected new holdings were created this would require 1.5 million Ha in the coming four years, which it linked with the amount of land in farms of over 100 Ha.[16] This logical connection was based on current thinking; the committee investigating *Osthilfe* expenditure had underlined that in return for public money the estates should render land for settlement. Indeed, Freiherr von Braun had estimated that if all the large estates in the east were split up no fewer than a 250,000 new holdings could result.[17] This was obviously a mouth-watering prospect for a movement engaged in the promotion of the peasantry.

The party as such had no animus, however, against large estates in particular. The March 1930 programme, the preamble to the EHG, Darré's 1933 guide-lines are all testimony to its general belief that a mixture of farm sizes corresponded to soil and climatic reality in the country. Darré's prejudice was really against the estate-owners in the east alone, since he held that unlike other parts of Germany the big farms there were the result of political and not economic developments. He chose a peasant meeting at Starkow (Pomerania) to air his views on the subject, to the general delight of the 'Blood and Soil' lobby. Darré pointed out that there was no reason why the eastern regions should not support a thriving peasantry, since history showed that they had done it before the Stein/Hardenberg laws: only these and not economics had started the depopulation of the area. Now the return of the peasants would be the first step in decentralisation of German industry. He referred to *Osthilfe* and the huge public subsidies given to estate-owners as a proof of the latter's inefficiency, and dismissed eastern estates 'from the standpoint of sober economic viability'.[18]

The speech caused a minor sensation even abroad; one foreign journal declared that 'War was declared upon the most powerful of the great land-owning barons.[19] In Germany itself the VB greeted it enthusiastically in a review article significantly entitled 'East Elbian Twilight'. The 'feudal economy' in the east would now disappear as a new aristocracy of 'Blood and Soil' arose from the peasantry. It seemed as though the death-knell of the Junkers had sounded, as after all both Darré and Rosenberg, the editor of the *Völkischer Beobachter*, were relatively in good standing in the party. Moreover, the former had already been at work behind the scenes, with a letter to the Chancellery enclosing statistics which proved, he felt, that the existence of the East Elbian estates could not be economically justified, which Hitler was said to have read with interest.[20]

Darré and his allies had other grievances which they could muster against the Junkers: in the first place, unlike peasants, they employed foreign labour, a point often made. Moreover, for the NSDAP they were politically unreliable; when Hugenberg left the government there was a wave of protest in East Prussia, which led to several arrests in the province.[21] This allegedly originated among local landowners, who were representing the DNVP leader as the only true friend of agriculture in the Cabinet. This may have been in Darré's mind when he told the landowners that they must accept the NSDAP and its policy of furthering the peasantry if they wished to continue with their estates: he suggested in his speech that they recognise the fact that the party had saved them from Marxism.

It seems very likely that Darré would have been pleased to have seen the back of the Junkers altogether; all he ever managed in practice was to take land from them in debt-relief schemes. This was a poor substitute for their annihilation. In an interview just after the NSDAP's accession he had spoken of the East Elbian owners throwing in their lot and 90 percent of their land with the new Germany, to become protectors of peasant homesteads as in the medieval era.[22] This proportion would have entailed their virtual disappearance as a socio-economic group. Moreover, as we have seen, solid political and economic reasons lay behind his desire, which Hitler might have been expected to share. If the Führer was not a genuine socialist he certainly was not a conventional conservative either. And yet once again the Junkers demonstrated their incredible talent for survival, as they had done under successive régimes.

It seems clear that they could not have outlived 'Blood and Soil' without powerful allies. The first of these was the President, whose sympathies for the whole landowning class were carefully exploited as before. Almost as soon as the new plans were mooted, two landowners hostile to Darré's settlement policy wrote to him: one described the policy of selling off large pieces of estates for settlement purposes as economic folly, and insisted that 100,000 new places could not be created by a wave of the hand. Hindenburg promptly passed these communications on to Hitler for his information. In addition, when a delegation of the Landowners Association was received by the Führer in early 1933 he mentioned large estates for which, he said, the *Reichsführer* SS had recently spoken up. As Himmler himself was a qualified agrarian expert, this advocacy was not unimportant, especially as Goering was apparently on the same side. In March 1933 he told the Pomeranian *Landbund* that settlement should not be carried out merely whilst existing forms of agriculture became ruined.[23] This may have been propaganda, of course, designed to reassure them of the NSDAP's good intentions, but ultimately as the Junkers were not proceeded against it may equally have been a genuine policy outline.

Perhaps an even more decisive factor was the traditional Junkers link with national defence, which continued during the lifetime of the Third Reich. Twenty noble families from the east contributed 160 members of the General Staff in the Second World War.[24] In 1937 an REM settlement official dealing with the difficulty of acquiring land for SS settlement wrote that 'The opinion has taken a firm footing in the *Wehrmacht* that the big estates are the best suppliers of the market.'[25]

There is no doubt that for rye and potatoes, staple foods in the Third Reich, there was an element of truth in this assumption. Even if 'best suppliers' was somewhat exaggerated, the big estates certainly played their part in food production. Because of this fact, and of the relationship between estate-owners and defence, the *Wehrmacht's* influence in helping to retain the larger farms may well have been considerable.

Equally the frontier question helped to shape the feeling that as its defence was traditionally under the leadership of a certain social class these people were vital to Germany. The Junkers had played on this prior to 1933. Von Gayl's letter to Hindenburg regarding Brüning's proposed settlement scheme referred to its possible deleterious effect on the 'powers of resistance of those circles who up until now have borne the will to national defence in face of the Poles'.[26] Hitler may equally have been of the same mind two years later.

Another aspect of their survival is even more interesting, as it bears directly upon Hitler's foreign policy, and puts the whole settlement question in a new light. In 1931 he denied in private any intention of carrying out large-scale internal colonisation in the east within the present frontiers; such projects were possible only when the land was available.[27] This was more or less in line with the March 1930 programme, which clearly stated that the acquisition of land for settlement was a task of foreign policy; obviously this does not necessarily imply the use of force, but equally it rules out the break-up of existing estates. There would be no point in recourse to that if fresh territory was to be obtained anyway. Ultimately this conviction of Hitler's may have been decisive in saving the Junkers.

So the 'Blood and Soil' enthusiasts were baulked by a combination of factors such as defence, foreign policy and agrarian economics. Consequently land supply for settlement was always restricted. There were still, of course, the provisions of the 1919 Act regarding the dissolution of the *Fideikommisse;* in 1937 some five-sixths of all new land acquired was made available from that one source alone. Similarly pieces of the big estates came on the market in respect of debt-settlement. In the year beginning 1 June 1933 about 80,000 Ha were acquired by this procedure.[28] However useful all this may have been it was manifestly in numerical terms a poor substitute for complete abolition of the East Elbian estates, especially in view of the increasing competition for space which settlement companies had to face from factories, roads and the *Wehrmacht.* By January 1938 an official report was bemoaning the shortage and pointing out that the sellers' market meant that only poor land was left for settlement purposes.[29] Another difficulty arose from the fact that the well-managed estates could not be bought out as they were needed for the struggle for self-sufficiency which ruled out expropriation, which as the report pointed out was the only alternative to purchase in the normal way.

The document is interesting for two reasons: first, inability to compete in bidding for land was admitted quite frankly, and suggests that despite 'Blood and Soil', little priority was afforded to settlement in the Third Reich. Secondly, there is the point that the efficient estates were needed for production: in other words, the purely practical considerations of saving foreign currency were recognised as taking precedence. Apart from the negation of Darré's guide-lines of 1933 contained in this premise, there is the additional point that the viable larger estates were accepted by the report as an economic necessity in the east, which provides additional evidence for their survival, despite Darré's own antipathy towards them.

The shortage had a catastrophic effect on would-be settlers; in the period 1934-36 some 20,000 certificates entitling the holder to a new farm under the settlement programme were issued, but only a little over half that number actually found a place. The situation did not improve, judging from a later REM letter stating that it was becoming steadily harder to deal with would-be settlers already accepted by the RNS and issued with a certificate.[30] The whole question of settler choice will be examined in another section but it has to be stressed here in passing that possession of a piece of paper entitling one to a new farm and actually acquiring the land were two very different things.

Settlement finance was always awkward, as the cost of land rose in the Third Reich; apart from competition between public bodies two other factors were involved, the debt-relief Act from Dr Hugenberg and the EHG. The first of these was clearly bound to have some effect, since when the large estates were heavily in debt owners were naturally more ready to sell strips of land than afterwards. The biggest single yearly increase in land prices before 1938 came between 1934 and 1935, or exactly at the time when Hugenberg's legislation started to take effect.[31]

As far as the *Erbhofgesetz* was concerned, this removed 55 percent of all agricultural land from the market: of course this land was not in itself suitable for settlement anyway, since there is no point in displacing one peasant merely to make room for another, but the limitations of land available on the market as a result of the EHG inevitably pushed up the price of what was left. By 1938 this had become so obvious that the RNS admitted it openly as a factor.[32] Thus to some extent an Act designed to complement the settlement programme was proving itself to be counter-productive.

Apart from land shortages the cost of building was another consideration, as it remained relatively high during the period 1933-38.[33] To this must be added new machinery and tools, plus stock: the first of these items alone was calculated to require RM 1,700 for the average holding. The final total was formidable, in 1933 the average cost of a new holding was RM 17,664 but five years later it was more than double. Since all prospective farmers were expected to pay a 10 percent deposit on the land before moving in, this created problems. In order to alleviate the financial burden various concessions were made, beginning with subsidies for the building costs, which varied according to the size of the holding, but on a total of RM 15,000 in 1938 a state subsidy of RM 8,800 was available. The balance of 90 percent on the land could be paid off on a long-term basis at 4 percent: this was staggered to give a transitional period for the first few years during which time interest rates were lower. State aid to purchase tools and machinery was also forthcoming in the form of special discounts: other financial assistance included reduced freight charges on building materials given in 1938. Families with four or more children got a special reduction in interest rates on the loan granted towards installation costs.[34]

Nevertheless finance remained a considerable problem: as early as 1934 the *Landesbauernschaft* in Bavaria was suggesting that more co-operative settlement should be carried out to reduce costs.[35] Although some activity along these lines has been recorded, it was slight. The *Artamanen* ran co-operative settlements on a small scale, in which everyone worked for board and lodging and pocket money. In East Prussia they carried out three projects.[36] But there is no doubt that the difficulty of finding capital seriously inhibited the programme, as was stressed by contemporary officialdom; in 1937 Willikens wrote of the capitalistic influence in settlement, and suggested that it must become possible for all wishing to become *Bauern* to do so: the obvious implication is that it still was not possible. The Rhineland LBF reported that SS men wishing to settle could not even afford the 10 percent deposit.[37] Darré himself expressed disappointment with the whole results at the Peasants Assembly in 1938, and conceded that settlement was not entirely free from the taint of capitalism: the way should be clear for every suitable young man 'even if he does not bring a penny of his own from home'.

A glance at settler selection in 1938 showed that in fact over two-thirds of all those accepted had RM 5,000 or more in the way of private resources and 40 percent had at least twice that sum.[38] As that was a considerable sum of money in the thirties, it seemed clear that poorer peasants' offspring or landworkers were excluded. NSDAP insistence upon a holding large enough to be viable played some part in this. This decision was not necessarily incorrect, but obviously the greater the size of the farm the more capital was required to get it, which at a time of rising land prices and building costs had especial point. The NSDAP programme thus wore a capitalistic air almost inevitably.

This had obvious implications for those landworkers who aspired to the status of *Bauer*: it was that type whom the party had specifically promised to aid in March 1930. The price of land was clearly important here, as von Rohr wrote in 1934, when he drew attention to the possible effects of the EHG on costs. Under the Republic the proportion of all new farmers formerly in the farmworker category multiplied four times between 1926 and 1930. Indeed, special financial encouragement had been granted in the latter year.[39] Now under the NSDAP the labourer found himself quite severely disadvantaged because of finance: this meant that even when he could get a place it was on average much smaller. So noticeable was this that a reader of the *NS Landpost* inquired for the reason, and the journal replied that landworkers were limited in management capability and also lacking in capital.[40] The point about ability may or may not have been true, but the remark about capital is surely revealing.

How inhibiting the lack of financial resources could be is exemplified in one scheme at Rierode (Hessen-Nassau). The holdings of 7.5 Ha each cost RM 8,900 for the buildings and RM 11-12,000 for land. On the first item a purchaser would have been obliged to pay approximately RM 3,600 and then over RM 1,000 as deposit for the land. But average annual wages in the area for a married farm labourer were only RM 1,314.[41] Opening the project Darré declared it to be 'a witness to the Führer's agrarian policy'. It was, but not in the way he meant it. Statistics are the final commentary on the landworkers' chances of becoming a peasant proprietor in the Third Reich; of 2,115 successful applicants for a place as a new settler in 1938 only about one-sixth were farm labourers.[42] Since there were at the time about two million in the country further comment would be superfluous.

One of the main reasons always advanced for settlement had been the need to guard the eastern areas for German culture. The limitations of the NSDAP's projects clearly militated against this, which would have been valid grounds for preferring quality to quantity in terms of new settlement. Consequently, as early as 1934 the RNS officials in the border regions were demanding a greater effort there. The Poles were apparently making great strides on their own side, including the dissolution of all estates of 200 Ha or over for the benefit of peasant holdings, an interesting commentary on the NSDAP programme. Hence, the Kurmark call to build a bulwark against the Poles who were 'pressing forward'.[43] At the beginning of 1936 the Ministry of Finance agreed to pay 20 percent of the cost of acquiring land in Polish border areas as an advance payment, and in 1937 extended this to the frontier areas abutting on Denmark and Czechoslovakia.[44] Party officials in Silesia and the Kurmark were also occupied with the question, as by July 1937 the areas were actually suffering a net loss in population through migration. Goering already knew the facts, and had placed contracts of a non-military nature there to try and stimulate industry. Significantly he did not even mention agricultural settlement, which suggests that not all leading members of the NSDAP shared Darré's views about its efficacy in promoting population shifts. In East Pomerania migration became so pronounced that a secret decree of 1937 actually forbade all movement to other regions.[45] The repopulation of the eastern part of Germany simply was not taking place at all.

This raised the whole question of frontier defence in a particularly acute form. In Upper Silesia the Polish minority was increasing so fast that fears were expressed that the whole region might be lost.[46] This strengthened the belief that the SS should be especially favoured in such areas, in order to build up racial/political strongpoints as a kind of focus for national defence. Himmler was in any case a keen advocate of eastern settlement and a close friend of Darré; the latter was, until February 1938, head of a department in the SS concerned with the vetting of recruits from a racial/biological standpoint (the RUSHA), as well as with the selection of SS settlers. There is considerable evidence for the belief that the two men were very close to each other until

Himmler apparently lost faith in Darré, hence the Minister's resignation from the RUSHA; but their ideas on race were undoubtedly very similar. What possibly caused the break in 1938 was divergent views about settlement.

By 1939 the SS had developed its own concept, based on the so-called 'defence farm' (*Wehrhof*). This would be about 30-50 Ha in extent and jointly run by two families, the heads of which would be SS men, racially and vocationally first-class; they would be peasants in the SS, in other words, not just any recruit. A conference took place on 24 February 1939 on the matter between a leading member of the RUSHA and Backe and Willikens as official representatives of the REM.[47] Both these latter laid emphasis on the need for the best quality peasants' sons from Lower Saxony; to underwrite their use the Ministry was prepared to contribute RM 500,000.

There was certainly no lack of suitable human material available for those projects. In Lower Saxony itself, Schleswig-Holstein and two other regions, one peasant son in every three joined the Waffen SS; as many as 90 percent of the original armed SS (*Verfügungstruppen*) had been brought up on the land. Incidentally on their side of the border the Poles had constructed settler villages from which anyone of German origin was banned. However racially-orientated the *Wehrhof* idea was, the NSDAP could legitimately claim it to be a counter-move only. Despite this, Darré rejected the whole scheme as impractical;[48] possibly he felt that too much emphasis was being laid on the purely military aspect of settlement. But equally he may have surmised that Himmler would use it as a lever to pry responsibility from his own hands, in other words, the SS would take over settlement altogether.

The presence of Herbert Backe at the meeting was not entirely coincidental; it was apparently under his influence that plans for agricultural settlement were produced at a conference in Hanover as early as November 1925, which 'read like a blueprint for later SS policy'.[49] Backe himself was born in Russia and came to Germany only after the First World War: he was in the RUSHA in 1933, becoming staff leader in 1937.[50] Because of his Russian connections, and his qualification as an agrarian adviser, he became regarded as an expert on the affairs of that country. This conceivably played a part in his nomination as Darré's successor in 1942, in so far as eastern colonisation may have been in Hitler's mind when he appointed him. In any event, his attendance, in conjunction with Darré's reluctance to promote the *Wehrhof,* implied that as far as the frontier defence aspect of settlement went, Backe was more likely to find favour at the top than was his chief. By 1939 the original Darré concept of a total re-orientation of the German economy had ceded place to frontier defence.

Statistics confirm this, as the yearly total of new holdings created diminished steadily, so that by 1937 the whole programme had almost ground to a halt. There were in the main three factors involved in this declaration. In the first place public interest appeared to contemporaries to be waning, so that demand for places fell off. Possibly the boom in industry had some influence here, but quite probably the party leadership had little real interest anyway, and gave the whole programme a relatively low priority. Defence was taking precedence, as one speaker admitted quite candidly when a new settlement scheme was opened in 1939.[51] Land procurement for military installations was the key point here.

Whatever the *Wehrmacht* took was in any case merely an additional loss. Once Darré had failed in his aim of dissolving the East Elbian estates virtually no other source was at hand. Debt-relief made landowners reluctant to sell and the EHG took away over half the area of agrarian land. The pressure of demand on supply simply pushed up prices to an unreasonable level, the opposite to the trend under Weimar. Under the Third Reich the price of land doubled, from RM 643 per Ha in 1932 to over RM 1,200 six years later. So between 1933 and 1938 the cost of an average holding complete with buildings and equipment more than doubled.[52] It is not then particularly surprising that applications fell to 2,605 in the latter year as compared to nearly 16,000 in 1934.[53] The NSDAP had always realised that land shortage would be a problem; when Willikens was asked before the party's accession where it could be found for settlement plans he declined to reply 'upon political grounds'.[54]

Finally, there was the question as to whether Hitler himself had ever had any interest in internal colonisation. It is notoriously easy, perhaps too easy, to display a continuity in his thinking about *Lebensraum,* from the time of *Mein Kampf* in 1924 onwards. The milestones along this by now much-documented route are his comments on 3 February 1933 about the 'ruthless Germanisation' of conquered territories in the east, the 1936 memorandum on the Four Year Plan with its emphasis on the widening of Germany's living-space, and the Hossbach protocol in November 1937. What runs as the Germans would say 'like a red thread' through these statements is the need to bring the frontiers of the country into balance with an expanding population. Here his private comments in 1931 assume a particular significance, as does the point about settlement in the March 1930 programme. There would be little need to worry about internal colonisation if one intended to acquire fresh territory anyway in the near future. Hitler's foreign policy will be assessed more fully in the conclusion, but it seems worth while pointing out here that it may well have been the biggest single cause of the surprisingly small scale of settlement projects after 1933.

The extent of the programme and its gradual decline accelerating post-1936, are evident from the figures given in Table 2.[55] In comparison with the record of the Republic (57,457 settlement places totalling 602,110 Ha) this achievement seems small beer. As the NSDAP had insisted on creating viable holdings, however, criticism of its results has to be tempered by comparing the average size of post-1933 holdings (about 16 Ha) with the much lower average of 10.5 Ha attained prior to that year. At least the new settlers had some chance of holding their heads above water. However limited the efforts of the NSDAP in quantitative terms it had reached its goal qualitatively.

Table 2

Year	No. of settlers	Area of holdings (Ha)
1933	4,914	60.3
1934	4,931	74.2
1935	3,905	68.3
1936	3,308	60.3
1937	1,894	37.6
1938	1,456	27.8
Total	20,408	328.5

This consideration applies equally to the individual capacity of the new men as it does to farm size; an examination of the criteria employed in their selection casts an interesting light on the people who became settlers in the Third Reich. Since Darré demanded in his guide-lines that only the best racial elements should be chosen, not merely the physical health and family background of the candidate were examined, but also his professional ability, character and political affiliations. Hence the comprehensive questionnaire which had to be completed, apart from a medical check and the consultations between the RNS and the local party leadership. Here again the KBF's testimony played a leading role, because the candidate had to be acceptable to him first.

All prospective settlers were required to be either married or engaged, to ensure that plenty of children would be available to carry on the settlement holding, most of which were *Erbhöfe.* In one case an engaged couple was only conditionally accepted solely in view of the bride's age, final permission was withheld until either a child was born or pregnancy could be proved.[56] One SS man was turned down because his wife had had a miscarriage with her third child.[57] Even the medical records of forebears were brought into question. In the case of an application from a married man, those of forty-two people in all were inspected before a permit could be issued. It was claimed that between 1933 and 1939 some two and a half million people's health was examined in respect of the 62,000 applications received.[58] Any kind of hereditary defect was sufficient to nullify a request, even evidence of suicide among the candidate's progenitors: the criteria became so sharply applied that complaints soon arrived from various parts of the country that capable, but still unmarried applicants were being too often rejected upon relatively slight grounds in this respect. Any person turned down was never given the reasons, incidentally, where health was the decisive factor, to avoid giving him an inferiority complex over his 'racial health'. Where a prospective bride failed to pass the test, however, the information was sometimes given to her fiancé, to enable him to avoid a biologically unsound marriage.[59] All candidates required the normal certificate attesting the absence of 'coloured or Jewish' blood in the family line since 1 January 1800, which was the normal EHG rule. This entailed a vast amount of local research, sometimes at the relevant *Landesbauernschaft*, which had a special Section G for racial questions (*Blutsfragen*).

Once the candidate's health had been checked the KBF could reply affirmatively to the questionnaire's point 'Is reproduction racially desirable? ' i.e. in biological terms. However, far more information was needed about character before a certificate could be issued. The local police chief was required to testify as to the candidate's record, his personal finance was scrutinized, as was the number of his illegitimate children, if any. More important still were his politics, the opening question for the KBF being 'Are the applicant and his wife/fiancée absolutely reliable in a political sense? ' War service had to given here, as well as any political affiliations, plus details of membership of any bodies connected to the NSDAP, such as the Hitler Youth. Even when a certificate was granted entitling the candidate to a holding, certain conditions for possible later withdrawal were set out, including the cessation of political reliability on the part of the aspirant.

Just how sensitive the RNS was in this respect was demonstrated in the case of a would-be settler in Oldenburg who, having got a certificate, was given permission to acquire an *Erbhof* near the border of Holland. Some ten months later, however, he was asked to surrender his settlement place and take a holding in Pomerania, which he refused. He then petitioned Celle on the grounds that he had been allowed a place originally by the *Anerbengericht.* The jurists took the matter up with the appropriate RNS authority: after some months the answer was given that the peasant concerned was politically unreliable, as he was a member of a certain sect which until 1936 had used Dutch for its church services. According to the RNS 'This fact alone justifies the rejection'. The point was further made that the certificate granted to any settler gave authority to acquire a holding in a general sense, which did not mean he could go anywhere he wished in particular; the right to decide whether he was suitable for that district still lay with the RNS.[60] The case illustrates NSDAP sensitivity over border areas, even in the west. Given this, the insistence on political reliability followed as a matter of course. Obviously there is no point in populating border districts with people who are less than fully patriotic, since their defence would then hardly be assured.

Once the strictness of the selection procedure is considered, it becomes obvious that it was no easy matter to acquire a permit to settle. One contemporary writer claimed that two-thirds of all applicants were rejected: this does not, however, square with the official statistics, which show a somewhat higher acceptance rate, approximately 50 percent for the period 1933-38 inclusive.[61] The deterrent effect of the rigorous examination must have played some part in keeping down the number of applicants, so that even the higher acceptance rate of the official figures does not necessarily imply that certificates were lightly granted. All available evidence seems to suggest that selection was meticulous and thorough in both medical and political fields.

Other aspects of the applicant reviewed included his professional ability, as he usually became a *Bauer* and was obliged therefore to meet the provisions of the EHG in terms of farm management. The inevitable consequence was a heavy bias in selection towards the more purely professional applicant who already had long agricultural experience. This is obvious from a study of the settler's former occupation: in 1937 as many as 95 percent had previously been employed in agriculture: nearly one-half were farmers already.[62] When the farmers' sons are added, two-thirds of all applicants were so comprehended. This proportion in fact rose steadily in the Third Reich since in 1934 only 48 percent of new settlers were either farmers or farmers' sons. Under the NSDAP there was an undoubted emphasis on sheer professionalism in comparison to the situation under the Republic. This took place despite the drop in the number of places for farmworkers already described, so that sociologically the average new settler after 1933 increasingly tended to be lower-middle rather than working-class, from a peasant background and not a rural labourer's (although landworkers formed nearly a quarter of new settlers in 1937).

A very clear pattern of geographical origin emerged over the period 1933-38 in that almost all settlers came from the same region in which they were granted holdings; in 1934 the proportion was officially calculated as three-quarters.[63] For eastern settlers this turned out eventually to be an under-estimate, since the final percentage for the period 1933-38 was as high as 82.6. Some of the remainder (5.3 percent) came from south and west Germany; at Darré's postwar trial it was alleged that these were mainly farmers from the latter areas uprooted by *Wehrmacht* land procurement. This seems, therefore, to have represented the only west to east movement, albeit involuntary: the original talk of a change in the pattern of population distribution in the Reich, so stressed in Darré's guide-lines, was never to materialise in any significant manner. This had always been the case, however, as in East Prussia in 1930 94 percent of all settlers were people stemming from that province; the Oberpräsident of the Rhine province said in 1934 that he could find records of only 344 families from his area who had gone east over the whole seven years from 1926 to 1933, and nearly half of them had taken non-agricultural employment there anyway.[64]

Clearly most people in the west did not wish to remove themselves to less hospitable regions. This was not the fault of the NSDAP, but the fact ought to have been considered by Darré when he wrote in the 1933 guide-lines that one-quarter of all places east of the Elbe should be given to applicants of southern and western origin. A closer acquaintance with reality would have shown him the lack of demand. It would appear in sum that Darré's hopes of a population shift, with agrarian settlement as the first step in industrial decentralisation for all Germany, was never more than a dream. Agrarian settlement provided the building industry with a turnover of RM 38 million in 1935, from its annual total of RM 7,200 million.[65] This surely demonstrates how small the scale of Darré's operations were, and how insignificant in terms of the economy as a whole. A genuine repopulation of the East Elbian regions by peasantry, even had land been available, would have required very little industry as agrarian purchasing power was relatively so limited: had the NSDAP really wished to decentralise the German economy, then industrial redeployment would have been the first step; this could never merely have followed in the wake of rural settlement.

Incidentally, the age-pattern of the new settlers is revealing, as in 1938 only 13 percent were under thirty, as compared to 70 percent in the range thirty to fifty, the remainder being older still. This seems to link with family size, since clearly applicants over thirty were more likely to have given tangible proof of their ability to produce large, healthy families than younger men could do, and men with large families were preferable. That this was so in practice can be seen from the relevant statistics, in that the number of settler families with three children sharply increased during the period 1933-38. The final impression is that family size and farming experience were given precedence over youth.

The general reluctance to move east must clearly have implications for Hitler's foreign policy. Why conquer more space if you cannot fill what you have? Of course, a greater supply of land would certainly have made it cheaper, so that victory would have removed one objection to participation. But it does rather seem that the judgement of a contemporary, that most of his fellow-countrymen were too 'spoiled' to want to move to the harsher conditions east of the Elbe, had a ring of truth.[66] From this it seems to follow that talk about *Lebensraum* was misplaced, except in so far as raw materials could be made available from occupied territories. But seen from the standpoint of settlement no one appeared to be interested. Perhaps ultimately a German politico-military hegemony over the native inhabitants would have been established, rather than outright annexation. Slovakia and the General-Gouvernement in Poland might have served as models in this respect.

There is, however, another possibility to which the Führer as a social engineer of some magnitude may conceivably have had recourse. He might simply have made compulsory population transfers. To a certain extent this was already implicit in the operations of Himmler's office established in October 1939, the RKFDV, whereby people outside the Reich of German descent were brought back within its borders. There is no apparent reason why the movement should not have been the other way if volunteers were lacking. There was a hint of this in Hitler's wartime remarks about the ease with which the Tyrolese could emigrate to the Crimea. 'All they have to do is to sail down one German river, the Danube, and they're there.' That they might not have wanted to go seemed to be irrelevant. The race took precedence over the individual in general, which could have applied to settlement as well as to anything else.

Alternatively he might have adopted a rather more Orwellian cast of thought and decided that some Slavs were less equally inferior than others. It is not inconceivable that a sort of officer-NCO-other rank pyramid could have been constructed in the east. At the top would have been the German overlords, and Poles, Ukranians and Russians tilling the soil at the bottom level, with more-favoured Slavs as intermediaries. Again, there was some precedent for this; Croats were always apparently more acceptable than Serbs, and Slovaks than Czechs. But whatever happened it is hard to see that the conquest of eastern territories would have evoked much response from German settlers as such. There is obviously the distinct possibility that Hitler had used *Lebensraum* merely as propaganda or that he was just out of touch with reality. A fuller discussion of this point is postponed to the Conclusion.

11

SELF-SUFFICIENCY IN THE THIRD REICH

The Contribution of German Agriculture

One of the problems which perpetually plagued the Third Reich was the balance of payments. Rearmament meant buying raw materials abroad which by definition could earn nothing by way of exports, as they were in effect consumed domestically. This was exacerbated by the chronic shortage of foreign currency which Hitler inherited from previous régimes. One way out of the difficulty was to try and cut down on other imports, such as food, by augmenting home production; German farmers thus had a vital part to play in aiding the armament and work-creation programmes.

The idea of saving money by producing more food at home had not altogether originated with the National Socialists. Increased domestic output had been a favourite theme throughout the twenties. In May 1926 the then Chancellor Luther told the Agricultural Council that Germany needed to be made as independent of foreign food as possible. Four years later the Prussian Minister of Agriculture underlined the same point by drawing attention to the current deficit in the balance of payments amounting to RM 3.2 billion: agricultural production must be stepped up to try and cover this.

German difficulties were obviously enhanced by the world economic crisis from 1929 onwards, which lowered purchasing power abroad and encouraged Protectionism, which eventually hit German exports. In 1930 the Americans fired the first shots in the campaign with heavy import tariffs in the Hawley-Smoot Act, and two years later Great Britain established Imperial Preference by the Ottawa Agreement, which foresaw in effect a kind of Commonwealth autarky. Italy was already engaged upon the *Battaglia di Grano* in an effort to make itself less dependent upon outside sources of food: the Soviet Union was by now committed to 'Socialism in one country'. There was a general move away from trade in the industrialised countries.

Consequently Germany was virtually pushed by circumstances into attempts at an even higher degree of self-sufficiency as witness the speeches of von Papen and von Braun to the Agricultural Council In June 1932. Both stressed the need for more output in view of currency shortages and foreign protectionism. Hitler made the same point in the following January to the NSDAP's agrarian cadre; he implied that in future more emphasis would be placed on the internal market in general rather than upon selling goods abroad. Darré was even more explicit: the 'export illusion' was over as far as the NSDAP was concerned. German attempts at conquest via international commerce had failed, henceforth the domestic market was to be predominant.[1]

Exactly how precarious the country's economic situation was can best be illustrated statistically. In 1933 Germany spent RM 3.6 billion on buying foodstuffs abroad at a time when its holding of foreign currency varied from RM 450 million in April 1933 to RM 280 million in the following June. The position was becoming absolutely desperate, especially because of the rapid fall in Germany's export surplus, which in the first few months of 1933 had shown a monthly average less than half that of the corresponding period in the previous year.[2] Added to this was the failure of international efforts to regulate trade and finance; in June 1933 the World Economic Conference broke down in London, making currency stabilisation impossible in the view of the German delegates who were attending it. German reaction was a measure limiting imports as far as possible, and protecting exports.[3] In other words, if others discriminated against German goods, self-defence was an absolute necessity, as international co-operation had failed. The position for the country was by now very serious indeed; she had to export to pay for food and raw materials and debt obligations, but was running so short of reserves that she scarcely had enough cash to buy raw materials in the first place.

Once rearmament and the creation of new jobs by means of state expenditure began to get under way the position deteriorated still further: funds devoted to arms alone rose from RM 0.7 billion in 1933 to nearly five times that amount in the following year.[4] The 1934 harvest was poor, so that there was a slight increase in food purchases abroad. More important still were the worsening terms of trade for Germany, primarily caused by a devaluation of the dollar. The net result of these three factors was a drop in currency and gold reserves which left the country almost totally bankrupt.[5]

It now became more necessary than ever to cut down on food imports, so that the maximum possible outlay on raw materials for the armament programme and for the export industry could be achieved. This resulted in the launching of the so-called Battle of Production *(Erzeugungsschlacht)* in November 1934 by Backe. He drew attention to commodities now being imported which could be cultivated at home, for instance, feed cake for cattle which from January to September had cost RM 182 million in foreign currency. At the same time Backe wrote on behalf of the REM to the *Reichsnährstand* calling on it to do its utmost to increase output.[6] The new drive was therefore launched not so much in response to previous ideological considerations about autarky as in consequence of current economic reality. The corollary of this seems to be that almost any government would have been obliged to carry out similar measures in view of Germany's plight.

The aim was always to maximise domestic output rather than to achieve absolute self-sufficiency, as Germany obviously could not cut herself off from world trade entirely. She had to export to some extent in order to pay for raw materials purchased abroad and in any case could scarcely take unilateral action to abrogate existing trade agreements. A classic example in this respect was Holland, a large-scale buyer of German industrial goods to which due consideration had to be given.

So when in January 1934 the two countries signed a new commercial treaty it was actually welcomed by Darré, although it foresaw more German food purchases. But as he pointed out, Holland had spent RM 630 million on buying German goods two years previously and was definitely a customer to be retained.[7] At first sight this seems like a reversion to the much-abused Weimar practice of giving exports priority over agriculture but now there was an important difference. The RNS had introduced guaranteed prices, so that foreign foodstuffs could be brought into the country without doing domestic prices any harm whatever, as imports would no longer automatically entail a fall. Darré therefore had no financial reasons for opposing the new bilateral accord.

Additionally he could see its solid political advantages for Germany. The Ottawa Agreement implied smaller British imports from Holland, so that Germany had a chance to step into the breach and pull it into its political and economic orbit. This gain justified in Darré's eyes any sacrifices for the German peasantry implicit in the new treaty. A row of similar agreements with countries such as Rumania, Hungary, Denmark and Bulgaria equally met with no objections whatever from the RNS.[8]

Equally, international trading in foodstuffs enabled Germany to get the benefit from any fall in commodity prices on the world market. In 1933, for example, there was a glut of grain in general; a system was employed whereby German cereals of a certain variety were exchanged for foreign grain *(La Plata)* to be used as fodder, by means of import/export certificates. Again, the prices for the imported type were fixed by the RNS.[9]

There was then, even in the RNS, complete recognition of the fact that agriculture had to serve the country's needs, it did not exist for its own sake alone. As a result foreign food would continue to be purchased to some extent whatever the results of the *Erzeugungsschlacht.* Even Backe had been aware of this before he launched the campaign.[10] This point has to be made in advance of any description of the *Erzeugungsschlacht* itself, otherwise continued purchases of foodstuffs abroad might make it appear that the new drive had been a failure. This was not necessarily the case, for the foregoing reasons.

The objectives in the campaign were best summed up by Darré himself; the essential aim was to reduce the space used for grain-cultivation but produce the same quantity as before by increasing yields: on the soil given up in this way fibre-bearing plants and fodder could be sown.

The Minister filled in the details by underlining that Germany was self-sufficient in grain but that gaps still existed in the cases of animal feeding-stuffs and fats, as well as in certain raw materials, hence the need for fibre-bearing plants as well.[11] There was nothing original in these objectives: in 1928 the need to increase the supply of domestic fats had been part of the emergency programme to aid agriculture. Dr Brandes had recommended more fibre-bearing plants to the Agricultural Council in 1932. As a result of poor domestic prices due to dumping from Russia, home cultivation had declined sharply prior to 1933:[12] the intention now was to further it again and save currency on raw material imports.

Backe had prepared a gigantic publicity campaign to put through the message to the rural population, involving an outlay of RM 357,000, with even a special magazine called *Erzeugungsschlacht.* The party joined in when Dr Goebbels ordered all speakers to support the new campaign at meetings.[13] The RNS itself used various propaganda weapons to urge on the producer to greater efforts. Apart from personal contact at meetings organised by the OBFs, these included loudspeakers in cars which toured the villages with a recorded conversation between two peasants, one against the drive, the other seeking to convince him of Germany's need for increased output. This was followed by a recorded speech by the relevant LBF. The whole considerable propaganda machine at the party's disposition was brought into play in order to win over the peasantry to the idea of a concerted drive in the national interest.

In passing one should emphasise that previous governments would have had great difficulty in even mounting it in the first place. One advantage of the RNS was that it represented a genuine, national, co-ordinating body for agriculture, which Weimar cabinets had never possessed. This explains partly why the RNS was so anxious to take over some administrative duties from the regional governments (see chapter 7).

It would seem that the RNS and the party needed all their undoubted powers of persuasion to convince the peasants in some cases that more output was needed. Price-falls of products such as hemp and flax had led to prejudice against them, and producers were still quite free to cultivate or not, according to personal choice. Darré quite specifically rejected recourse to any compulsion. He felt that the production battle could only be won if appeal was made to the peasant's honour and voluntary spirit; to try and force him was useless.[14] However, some social pressure was undoubtedly applied despite the absence of legal measures. Only a few months after the campaign began one OBF in the Rhineland drew up a list of the peasants in his village who did not habitually attend the special *Erzeugungsschlacht* meetings.[15] It would be surprising if the wish to bring some kind of moral pressure was not behind this. But apparently sanctions stopped there, short of actual legal enforcement.

Since Germany could scarcely increase food output without more feeding-stuffs, currency allocations for these purchases were evidently essential to the success of the whole operation. Eventually it was hoped to cover the need for protein-rich fodder from domestic output but initially most of this came from foreign sources. Unless the necessary currency was put at the disposal of the RNS and the Ministry of Agriculture to continue those imports then the *Erzeugungsschlacht* was doomed from the very beginning.

Unfortunately, this matter became the centre of friction between Darré and Schacht. For his part Schacht had no easy task in trying to finance work-creation programmes when German reserves were so low. As he later wrote, it was no bed of roses being Germany's Economics Minister at a time when every Mark had to be watched carefully.[16] He was almost trying to square the economic circle. On the other hand Darré had a strong case too. There was simply no point in expecting him to cover the Fats Gap if he was not allowed to buy oilcake and other fodder abroad as an ineluctable preliminary.

Dissension between the two men had already developed, at first over the question of grain purchases abroad. The doctor had no objection to acquiring cereals from countries with which Germany had exchange-agreements, but he contested all those cases where payment had to be made in foreign currency. The Reichsbank wrote to the REM nine times in twelve months from April 1933 onwards about the need to husband the scanty reserves in this connection.[17] Schacht eventually went over Darré's head to the Chancellery and managed to get the export of flour stopped, if nothing else.

By 1934 the position was deteriorating still further for Germany: an additional factor here to those previously given was the opportunities for lucrative contracts on the home market which was apparently dampening the zeal of businessmen for exports. Eventually Hitler had to issue a special decree to ensure that only firms with a certain sales performance abroad could be allowed public contracts in Germany. This evidently proved effective but until it came out exports were falling behind expectations.

Then to make things still worse came the poor harvest of 1934-35 which naturally increased food imports. Moreover some ill-judged measures by the RNS, still relatively inexperienced, exacerbated the situation. The peasants were initially asked in 1934 to surrender 70 percent of their rye harvest at once in exchange for imported barley, to be used as animal feeding-stuffs, especially for pigs. Many poorer people used dripping as a fats substitute, and plenty of pigs meant an adequate supply, which would reduce pressure on the demand for butter and margarine. However well-conceived theoretically, the idea did not work: the rye which was delivered by the peasants was badly handled at the depots and in some cases it had to be handed out again as animal feeding-stuff. In some regions where rye was normally used for fodder anyway the peasants saw no point in surrendering one form of feeding-stuffs in order to acquire another, and after objections the original RNS quota of 165,000 tons of rye from Schleswig-Holstein had to be reduced to 100,000 tons.[18] All in all, reduced grain harvests and RNS inexperience produced a grave shortage of domestic fodder; imports of both barley and maize were greater in 1934 than in the previous year to cover the gap, but this was not in itself sufficient, and the stocks of both cattle and pigs were smaller in 1935 than they had been two years before.[19]

The *Erzeugungsschlacht* had got off to a terrible start, and Darré had recourse to new measures to distribute limited supplies as effectively as possible. The production of whipped cream was restricted to try and conserve milk, and butter-producing districts had to send 10 percent of their output to central depots for allocation to less-favoured areas. It was decided to impose a quota on all pig-slaughterers equal to 70 percent of the swine killed by them in the previous year; this would prevent any panic slaughter as a result of which the current shortage might be solved only at the expense of an even worse one in the future.

These measures did not work; the combination of this failure with currency limitations, produced a really quite chaotic situation in November/December 1935, when current reports read as though emanating from a besieged country.[20] A vicious circle set in, where shortages producing ineffective regulations, then transgressions, were followed by a worse shortage and even more regulations. As an illustration, the quota on slaughtering was a dismal failure, because no one had apparently been giving the right figures previously for the number of animals killed; this casts an interesting light on RNS marketing in general. It would appear that rules had been evaded as slaughterers had evidently been killing more pigs than they admitted in order to sell the others at above the fixed price. As the 70 percent quota now obviously applied to official returns given, and not to the actual numbers killed, the total of pigs which could now legally be slaughtered was far less than in the previous two years. Hence the 'wide overstepping' of demand over supply mentioned in this particular report: in desperation the pig-owners began to evade the quota, which as a result of the closer supervision now employed, led to arrests. This was by no means confined to one region; in Bavaria the Minister of the Interior authorised the use of concentration camps for anyone caught rigging the food market in 1935-36.[21] In Lower Saxony, however, the widespread use of arrests was not advocated in some party circles, as it only led to bad feeling on the land.

So serious was the situation by early winter 1935 that the press began referring to fats shortages quite openly, as in the DAF magazine which ran two articles on the question. An over-simplified explanation put all the blame on to lower imports. But the accounts conceded in effect that some sections of the population were markedly dissatisfied with the position.[22]

Understandably the Schacht-Darré battle took on a new sharpness. The Economics Minister accused the REM of repeated failures in its calculations regarding currency needed for imports. Hitler was appealed to and nominated Goering as mediator, who sanctioned more oilseed imports. This intervention was not sufficient to halt the strife and in March 1936 Darré wrote to Hitler, Goering and Schacht warning them of the consequences if more currency was not granted for food purchases. This drew a stinging reply from the Minister of Economics. He absolved himself of any responsibility in the matter and told Darré that he must ensure a level of agrarian output equal to the one he had inherited. Since stocks of farm animals had dropped and grain yields had been poor, this observation, however blunt, had some factual basis. Schacht concluded with the exasperated remark that he could not actually conjure money out of a hat *('Ich bin nämlich kein Dukatenmännchen').* Despite this, Darré's appeal to Hitler bore fruit and the Führer allocated him a further RM 60 million.[23]

Ultimately when the whole matter is considered in the light of Germany's overall economic position the mutual recriminations seem to have been pointless, at least in one respect. Both Darré and Schacht were parties to building up the *Wehrmacht* and to solving unemployment by public works. That was bound to strain the economy since foreign exchange was in short supply: the poor harvests of 1934-35 were the culminating factor. As early as August of the latter year Schacht was describing the fats situation as precarious: imports would have to be further limited in future, rather than priority being given to them over raw materials for industry.[24] In March 1936 he drew up a further memorandum on the current state of affairs. Rearmament as envisaged would need a 25 percent increase in exports to counterbalance raw material purchases, but he could not at the moment foresee more than 10 percent being attained.[25]

Coincidentally it transpired that Hitler was actually contemplating an even quicker *Wehrmacht* build-up. Arms expenditure for 1936 was already as much as thirteen times as great as three years before.[26] Moreover, raw material prices were moving upwards against Germany, by 10 percent in 1935-36, which clearly justified Schacht's pessimism. On top of all that it turned out that Backe had over-estimated the possible increase in domestic agrarian output in October 1934 when the *Erzeugungsschlacht* was launched. The REM therefore wished to spend even larger sums on imports in the second half of 1936 than in the first. Unfortunately raw material stocks for industry were low, only 1-2 months supply being available, as compared to 5-6 months in early 1934.[27] There were no savings likely to be made in that respect in the immediate future, even at the present armaments level. It appears hardly surprising then that Schacht should have asked for a slowing-down in armaments production in May 1936, at a meeting of the Exchange and Raw Materials Committee, of which Goering was now in charge to get better results. By summer 1936 Hitler was moved to far-reaching measures, since he was apparently deaf to Schacht's appeal to slow down the *Wehrmacht* build-up. The initial step was a memorandum.

Exactly when the Führer put his thoughts on paper has never been determined, but it would appear to have been in late summer 1936, since the memorandum was first read out to the Exchange and Raw Materials Committee on 4 September 1936.[28] The document began with a political survey, in particular the threat which Soviet Communism posed to Germany: 'over and against this danger all other considerations must step back as totally unimportant'. The Führer then dealt with economic affairs, pointing out that self-sufficiency was easier to achieve with six million unemployed than would otherwise be the case, since clearly re-employment increases people's purchasing power and therefore their consumption. As the unemployment problem was rapidly being solved there was now no hope of achieving full autarky in foodstuffs from Germany's present soil. This exactly parallels the earlier view expressed by Hitler, 'that increasing needs always militate against self-sufficiency in food supplies'.[29] The inability of the RNS to close the gap in this respect seem thus to have buttressed the Führer's original logical deduction with actual empirical evidence. In fact, the country was 81 percent self-sufficient in 1936, a slight improvement on the situation obtaining four years previously.[30] But Hitler's conclusion was none the less justified in view of the current position; the *Erzeugungsschlacht,* partly due to two bad harvests, had been a relative failure. If the Führer wished to expand the armed forces still further the tempo of the agricultural drive would have to be speeded up.

Hitler now turned to the alternatives; a trade drive abroad could theoretically solve the problem of rising living-standards at home, but this could only be achieved at the expense of armaments (which is virtually what Schacht had maintained) and the claims of the latter could not be deferred under the present circumstances, due to the Soviet danger. The Führer eliminated international trade once again, in other words, on politico-economic grounds: his summary was that 'The definitive solution lies in a widening of our living-space'. The memorandum concluded that 'The German Army must be capable of operations in four years: the German economy must be ready for war in four years.'

The document seems to suggest that Hitler's views had not changed in any way in the course of the twelve years since his autobiography had been dictated. Autarky is desirable but it cannot be achieved under existing conditions, therefore let us change our conditions, may be described as the Hitlerian syllogism which summed up his economic outlook. It is unlikely that the failure of the RNS to achieve virtual independence in food supplies contributed much to this thesis, since Hitler's mind had apparently always worked that way: but the situation in the summer of 1936 may well have confirmed his views that self-sufficiency inside the existing frontiers would be impossible. Even if the *Erzeugungsschlacht* had been more successful it seems likely that Hitler would still have proceeded with his plans because of the existence of Communism in the Soviet Union.

On 18 October a directive entrusted the execution of the new programme to Goering, and the economy was divided into six groups of which agriculture was the fourth. The goals assigned to the land were the direction of consumption, the battle against waste, planned utilisation of available labour, and the scientifically organised exploitation of untapped resources in the soil. To stress the seriousness of the position in general the death penalty was now introduced for 'economic sabotage', a measure enacted at 'the wish of the Führer'.[31] Herbert Backe was given the job of co-ordinating agrarian policy and industrial planning, for which he would be responsible to Goering himself and not to

Darré. The total resources allocated to agriculture in investment terms amounted to RM 1.5 billion for the period 1936-40, or 16 percent of the total outlay envisaged for the Plan as a whole. The importance of increasing domestic fodder supplies was shown by the RM 152 million devoted to this alone; this was chiefly to enable the fats gap to be closed.[32] The new Plan represented greater investment in agriculture compared to previous figures, since from 1933 to 1936 inclusive the total expenditure by the REM had amounted to only about half that now earmarked under the Plan. The Ministry continued to dispose of its own funds, of course, which also rose sharply after 1936.[33]

Initial measures to improve output were announced by Goering in March 1937; these included a cut in the price of fertiliser and large sums for soil-improvement and the winning of new land. The price of rye was increased to help bread supplies, the idea being to discourage its use as animal feeding-stuff (the fodder gap would be met by increased beet, potato and clover, etc. crops). There was to be better housing for landworkers and subsidies to assist mechanisation. To Georing's factual announcements was added an emotional appeal from Darré: 'I call you, German land people, to a competition in production.'

In East Prussia alone Erich Koch believed an annual increase in food to the value of RM 100 million to be possible, to which better drainage and other soil-improvement methods would contribute a half and more fertiliser another quarter.[34] At a national level it was intended to raise fibre output from domestic sources from 15,000 to 80,000 tons. No source of food was to be neglected in the new drive, particularly where fats were concerned; it was even announced that two new whalers were under construction, as whale oil could be used in margarine production. The laying capacity of hens was to be raised and the slogan for poultry-keepers was to be 'Produce your own feeding-stuffs': even bee-keepers were given new targets to attain. Goats were to produce more milk and by improved breeding, etc. the supply of rabbit-meat was to be increased.[35] In sum, the *Erzeugungsschlacht* was to be greatly intensified.

This meant that no considerations about private property could be allowed to interfere with output. As one internal directive put it: 'In the interests of production . . . every bit of German land, insofar as it is agriculturally utilisable, must be cultivated in the best possible fashion.'[36] The clause in the EHG which empowered the RNS to remove the less efficient *Bauer* from the management of his holding was in effect extended to all farms in the country. A new measure authorised the RNS to place badly-run properties under a trustee or to lease them compulsorily to a person experienced in agriculture. In presenting this bill to the Cabinet Darré requested that not too frequent a use be made of the new legislation: its presence should be a deterrent in itself.[37] The measure allowed in any case for a warning to be employed if necessary, rather than actual eviction.

The law was none the less applied in some cases, as in Hagen (Lower Saxony) where the OBF felt that a local farmer was no longer capable, due to age: the KBF took the case up and told the man concerned to lease the best part of his land or a formal proposal for compulsory leasing would be applied for.[38] One *Bauernführer* wrote to the owner of 15 Morgen lying idle asking if he was prepared to let it, and giving the name of a neighbouring farmer as a likely candidate. But the RNS was not always quite so determined, judging from other cases on record.[39] Darré's wish that the mere existence of the regulation should be sufficient was generally observed.

Deliveries of grain to the RNS were now tightened-up as well; henceforth all production had to be officially handed over, instead of merely that amount for which the organisation would give a fixed price. The grain Federation was empowered to make exceptions to the new rule where it saw fit. Darré ordered the *Bauernführer* to draw up a list of all producers who pledged themselves not merely to be punctual with deliveries but who were also prepared to exceed their quotas: a committee would be appointed in every parish to check on whether the duty was being complied with. Here again, it does not seem as though the matter was always prosecuted with absolute vigour, since some months later the deputy RBF, Behrens, found it necessary to appeal to farmers to observe the decrees.

A much closer control was observed over the distribution and use of fodder under a decree of July 1937; it was forbidden now, under pain of a fine of RM 100,000, to feed grain suitable for bread to poultry. Similarly, a control office took over the distribution of pig feeding-stuffs which it handed out to farmers direct. The prohibition on bread grain was expected to save two million tons of rye and a half a million tons of wheat for human consumption yearly: the law seems to have been enforced since in Detmold a miller was arrested for using rye as grist fodder.[40]

Supervision of farming activities in general was facilitated from 1936 onwards by the introduction of a record card for every holding over 5 Ha in size. This required exhaustive details to be rendered by the occupier, a copy being held by the KBF. These were another step in the closer control of individual enterprises, and eventually became an important tool in the wartime economy of Germany. Neutral in themselves, the cards symbolised that 1936 was a milestone in German agriculture. Altogether over two million holdings containing some 90 percent of all agricultural land in Germany were comprehended in the new system.[41]

All in all, there was a noticeable increase in pressure on the agrarian sector from 1936 onwards, in which the Four Year Plan played a definite part, with an even greater emphasis than before on production and controlled distribution. It is probably not an exaggeration to say that its initiation saw another step towards a wartime economy in Germany three years prior to the outbreak of actual hostilities.

However, too much should not be made of this, as to some extent the formation of the RNS as such had introduced central control for agriculture. The Four Year Plan represented for the agrarian sector a closer degree of supervision, rather than any new principle. In any case, there was often a gap between the regulations in theory and their actual application, as in the case of delivery quotas or removal of the farmer from management. Darré had virtually asked for this in Cabinet, as already mentioned. As with the EHG, the authorities seemed ready to compromise with the agrarian sector in these respects too. In the final analysis this flexibility may have been simply a recognition of the fact that the farmers were indispensable to the economy, so that more dictatorial methods would have been counter-productive.

From a purely technical standpoint the most important changes introduced under the Plan were the cuts in fertiliser prices and the increased investment for agriculture. To persuade the manufacturers to accept the first was no easy task according to an RNS official concerned with the negotiations.[42] That the move was right, however, can be seen from Table 3 which shows the rise in fertiliser use afterwards.

Clearly there had been a trend towards a more extensive use even prior to the cuts, but they obviously provided an additional stimulus. This was important, as Germany was relatively backward in this respect. In 1934 it was estimated that another 100,000 tons annually could have yielded two million extra tons of grain.[43] Moreover, peasant leaders themselves knew this and had constantly complained about the

Table 3. Fertiliser use by type (000 tons)[44]

	Nitrogen	Phosphate	Potash	Total
1932-33	353	470	717	1,540
1935-36	490	636	949	2,075
1938-39	718	749	1,254	2,721

high price of fertiliser which militated against its greater employment. The *Deutsche Bauernschaft* pleaded in vain for a price-reduction of one-quarter. Here, with the new cuts, was another reason for peasants to be satisfied with the régime.

Increased funds made available under the Four Year Plan stimulated land reclamation and improvements in general. The Labour Corps was often utilised on such schemes as the Adolf Hitler Koog on the coast of Schleswig-Holstein, although to obtain 1,350 Ha of new land cost RM 3.7 million even with the use of cheap labour. This was a ruinously expensive way of acquiring soil, however much the press publicised such schemes. But the true cost was never revealed. Improving waste land was cheaper than reclamation, of course, but even there in view of the marginal gains made in proportion to Germany's total agrarian area it was hardly an economic policy.[45]

In effect such projects were the agrarian equivalent of the scheme to use German iron ore at enormous expense rather than purchase higher grade material from Sweden with foreign currency. In the short term this could be justified in view of the objectives in the Four Year Plan, namely, to build up a war economy as rapidly as possible. Politics simply took precedence over economic theory, which applied to agriculture as much as to anything else. However costly land reclamation was to the nation it enabled precious reserves to be devoted to raw materials for armaments rather than being spent on food from abroad. Well-publicised schemes were also a feather in the cap for the régime, a political bonus as it were.

Not all capital investment in agriculture was so relatively unproductive in the long run, in any case. Poor soil capable of yielding vegetables only was converted to ploughland for other crops, or to grazing: some 280,000 Ha were treated in this way. Pasturage could also be bettered with the aid of subsidies to individual farmers to the tune of RM 80/100 per Ha. For ploughland long-term credits were made available only to public bodies such as co-operatives. But a vast amount needed to be done: it was estimated in 1939 that a million Ha stood in need of flood protection, four million required drainage and another seven million irrigation. For the period 1937-40 RM 1,000 million was earmarked for land improvement in general, but the sheer size of the problem, which the NSDAP had inherited, defied any real short-term solution.[46]

One way of increasing output from existing land was a greater use of machinery. The scattered nature of many holdings and their relatively small size made the mechanisation of agriculture in Germany difficult, another factor being conservatism among the peasants. It was said that 'The farmer buys no machines, machines are sold to him.' In the Kassel district in 1929 only 4.7 percent of all holdings had power-driven machinery and only one-quarter had machines of any kind.[47]

Basically Darré was in favour of mechanisation, without which he believed production increases would be impossible: on the other hand he was to a certain extent the prisoner of his own ideology in the matter, since he rejected collectivisation on purely racial grounds, although it was easier to manage, as he admitted.[48] This attitude implied a continued multiplicity of holdings, no aid to mechanisation. The government did facilitate it as far as possible by giving subsidies, especially to new settlers, for machinery purchase and by encouraging co-operative buying. Higher incomes for food-producers obviously played some part here, and in 1938-39 German farmers were spending far more on acquiring new machines than they had six years previously. But a comparison between tractor use in the Third Reich and that in contemporary Britain puts the German backwardness into sharp relief. Whereas Britain had one tractor per 130 Ha, Germany had only one for every 388.[49] Of course, this was again an inherited situation for the NSDAP and a large-scale output was planned from 1940 to improve matters. But the party's ideological commitment to the retention of the peasantry always militated against a really rational and mechanised agriculture.

Moreover, after 1935 land reclamation or improvement was balanced by the loss entailed by military expansion. Two separate measures were introduced in 1935 regarding land procurement, one a general law, the other relating solely to acquisition by the *Wehrmacht*.[50] When the latter was accepted by the Cabinet Darré was present and raised no objections. At his postwar trial it was maintained that 'every Ha of land which was taken away from the farmer was fought for'; on this occasion, however, no resistance from the Minister was forthcoming. Presumably he could at least have said something, since after all the peasants were protesting themselves at the time.[51]

As for the second bill, this was introduced in Cabinet by Darré himself and instituted a general office of land procurement under the Führer. This could have been used for settlement purposes, for the military or for forest and road requirements. Some land was eventually lost to the autobahn programme but the scale does not seem to have been large judging from records of *Erbhof* courts transactions for land sales. It is in any case difficult to see how a large-scale road programme can be executed without displacing someone.

There is no doubt that the *Wehrmacht* took most of the land lost, its claim for the Siegfried Line alone being 200,000 Ha according to Darré at the sixth Peasants Assembly in 1938, from a total figure of 370,000 Ha lost. Another contemporary estimate was as high as 650,000 Ha up to 1937.[52] This was a generalisation as the Weser-Ems area actually had more land in 1937 than four years previously due to reclamation.[53] How the farmer was affected by procurement depended on where he lived, but this does not alter the fact that the country as a whole disposed of less agricultural land as the result of the *Wehrmacht* build-up. The tempo quickened so that by early 1939 the RNS was asking all areas if they had begun to compile a *Heimatbuch* like the *Landesbauernschaft* in Lower Saxony containing records of all displaced peasants. The occasion for the loss of their original holding was plain; 'on account of the creation of industrial facilities and manoeuvre areas for troops a greater number of peasants must be resettled'.[54] Ironically, Willikens actually lost his own farm later, due to expansion of the Hermann Goering Works; the holding had been in the family for three centuries.[55] Land loss cost the country dearly, which has to be seen as counter-acting efforts in other directions at augmenting domestic output. This is peculiarly ironic, as the post-1936 *Erzeugungsschlacht* was designed to assist the military build-up; now the latter was impeding its attempts at doing so.

However, from available statistics it appears that the campaign was in general a success. In 1938-39 production on the land was one-fifth greater than it had been ten years previously, half of this having been achieved since 1935-36;[56] clearly the Four Year Plan had speeded up the tempo. Consequently Germany's degree of self-sufficiency was markedly heightened, food imports fell and foreign currency was saved, which was the main object of the exercise. Moreover, as well as by an increase of about two millions in the total population in the thirties, the food situation was potentially threatened by an improvement in qualitative terms. In other words, Germans were on the whole living better, as Hitler pointed out in his memorandum prior to the Four Year Plan. His claims are substantiated by independent statistics (see page 181). These show conclusively that even post-1936 the standard of living appeared to be rising, in that the consumption of foodstuffs per head was greater than in previous years. So however much Goering may have spoken of 'guns before butter' such a choice was never really presented to the German public, at least before the war. Clearly the success of the RNS in boosting domestic output played a major role here; had it not happened, then Hitler would have been faced with an ineluctable decision, less guns or less butter. The first was politically unacceptable, the second fraught with all kinds of potential trouble for the régime. That he never had to take it was due to the *Erzeugungsschlacht;* thus when Darré said in public in 1939 that the campaign had won Germany the chance to follow an independent foreign policy he spoke no more than the truth. In this fact lay the main justification, and success, of the production battle from a National Socialist standpoint.

Approval none the less has to be qualified because of the relative failure in respect of fats, which continued to plague the economy. Even in 1938 domestic output equalled only 57 percent of consumption, little more than four years previously. Backe attributed this squarely to the shortage of feeding-stuffs.[57] Whereas average annual domestic fodder harvests for the period 1928-32 amounted to 9.8 million tons, for 1936-39 the corresponding figure was 10.9 million tons. The improvement just was not large enough, despite the frantic appeals to the peasantry made by Darré at the Peasants Assembly in both 1937 and 1938. Shortages in domestic fodder, when coupled with scarcity of foreign currency, inevitably restricted the stocks of pigs and cattle and therefore of fats. This shortfall had been the centre of the emergency programme in 1928 and ten years later there was little change.

This partly accounts for the fluctuations in currency saving which may be attributed to the *Erzeugungsschlacht,* when food purchases abroad are expressed as a percentage of total imports. In 1930 this proportion had been 40.7 percent from which it declined to 35.5

percent in 1936. Two years later, however, it was up to 38.8 percent.[58] Imports of vegetable products in general (grain, fruits, fodder and vegetables) declined far more sharply than those of livestock and animal products. What makes the significance of this difficult to assess, however, is the existence of the various bilateral trade treaties which entailed buying foreign food in exchange for export outlets for Germany. This was particularly so after 1936 when Dr Schacht toured south-eastern Europe in the hope of meeting Germany's desperate need for new markets. The sequel was a series of treaties accepting German goods against mainly agrarian produce from the Balkans. The obvious outcome was a rise in imports, which implies that figures for food purchases abroad are not an absolutely certain criterion for the success or failure of the *Erzeugungsschlacht.* Maize imports, for example, rose from 338,000 tons in 1936 to nearly two million in 1938, due to the trade with the Balkans.[59] Statistics show conclusively how German commerce literally re-orientated itself in the Third Reich. The six south-eastern countries sent Germany 4.6 percent of her imports in 1929, but 11.9 percent by 1938. Exports shows a roughly similar growth. As items such as grain, eggs and soyabeans (for fodder) figure so largely among the post-1936 transactions it would seem that analysis of the *Erzeugungsschlacht* needs to be modified correspondingly, especially as bilateral treaties with the Netherlands and Denmark had been signed as early as 1934.

Under these circumstances a statistical summary of the *Erzeugungsschlacht* is perhaps the most appropriate way of illustrating its achievements (see Tables 4-11).[60]

Table 4. German agricultural production and stock holding by commodity (animals in millions, crop products in million tons)

Year	Commodity								
	Wheat	Rye	Barley	Oats	Potatoes	Beet sugar	Sheep	Cattle	Pigs
1930	3.8	7.7	2.9	5.7	47.1	14.9	3.5	18.5	23.6
1932	5.0	8.4	3.2	6.7	47.0	7.9	3.4	19.2	23.0
1933	5.6	8.7	3.5	7.0	44.1	8.6	3.4	19.8	24.0
1934	4.6	7.6	3.2	5.5	46.8	10.4	3.5	19.3	23.3
1935	4.7	7.5	3.4	5.4	41.0	10.6	3.9	18.9	22.8
1936	4.4	7.4	3.4	5.6	46.3	12.1	4.3	20.1	25.9
1937	4.5	6.9	3.6	5.9	55.3	15.7	4.3	20.5	23.8
1938	5.6	8.6	4.2	6.4	50.9	15.6	N/A	19.4	23.6

***Table 5. Land sown with fodder and fibre-bearing plants (000 Ha)**

Plant type	Year				
	1928	1933	1934	1936	1938
Rape	16.8	5.2	26.7	54.6	61.9
Flax	14.5	4.9	8.8	44.1	44.9
Hemp	0.8	0.2	0.4	5.6	12.7
Corn maize	2.1	3.7	16.3	19.3	65.8
Sugar beet	454.0	304.0	357.0	289.0	502.0
Winter barley	183.0	271.0	307.0	436.0	517.0
Lucerne	285.0	315.0	319.0	404.0	412.0
Lupin (fodder)	46.0	53.2	57.3	81.7	101.6

Table 6. Yield per Ha in selected products in double Zentners (1 Zentner = 1 cwt) in German agriculture

Product	Average yield 1929-33	1938 actual harvest
Rye	17.4	19.8
Wheat	21.5	23.6
Barley	21.6	22.2
Potatoes	156.1	182.2
Sugar beet	283.1	333.4

Table 7. Degree of self-sufficiency for Germany in selected foodstuffs (%)

Product	Year		
	1927/8	1933/4	1938/9
Bread grain	79	99	115
Potatoes	96	100	100
Vegetables	84	90	91
Sugar	100	99	101
Meat	91	98	97
Eggs	64	80	82
Fats	44	53	57
All food together	68	80	83

**Tables 5-11 are reprinted from 'Autarkiepolitik im Dritten Reich', D. Petzina (1968) by permission of Deutsche Verlags-Anstalt, Stuttgart.*

Table 8. German foodstuff imports in value (RM 000 million)

1929	1933	1934	1935	1936	1937	1938	1939
5.5	3.6	3.7	3.2	3.1	4.1	5.0	4.4

Table 9. German foodstuff imports by value and product as a percentage of total imports of all types of goods

Product	Year							
	1929	1930	1932	1933	1934	1936	1937	1938
Livestock	1.1	1.1	0.7	0.7	0.7	2.3	2.0	2.1
Animal products	11.5	12.6	12.7	10.3	8.7	10.5	8.8	8.8
Grain, fruits and vegetables	21.9	20.9	25.4	20.7	18.6	15.9	20.8	21.5
Coffee, wine and tobacco	5.5	6.1	6.9	7.1	6.7	6.8	5.9	6.4
Together	40.0	40.7	45.7	38.8	34.7	35.5	37.5	38.8

Table 10. Grain imports by quantity (000 metric tons)

Type	Year					
	1929	1930	1933	1934	1937	1938
Wheat	2140.8	1197.2	770.3	647.0	1219.0	1267.7
Barley	1765.8	1522.9	235.3	552.0	241.8	456.4
Maize	669.0	651.3	254.1	338.3	2158.9	1895.4
Rye	144.0	59.0	238.0	53.0	181.0	N/A

Table 11. Consumption of various foodstuffs (kilos per capita yearly)

Item	Year				
	1933	1934	1935	1937	1938
Butter	7.4	7.4	7.5	8.9	8.8
Meat	42.0	45.4	44.2	47.2	48.6
Milk	104.3	106.7	107.7	109.0	112.0
Eggs*	119.1	118.0	112.3	124.0	124.0
Cheese	6.1	5.8	5.5	5.4	5.5
Potatoes	187.1	179.9	173.5	174.0	182.9
White flour	46.5	49.2	51.1	54.3	51.9
Rye flour	54.0	52.6	53.5	55.2	53.0

*Total number consumed.

12

RURAL MIGRATION BETWEEN 1933 AND 1939

Its Extent and Causes, and the Consequences for the *Erzeugungsschlacht*

Despite the relative success of the drive for more output on the land it might have been even better but for rural migration. This was again a problem of long standing in Germany, which the NSDAP had done nothing to create. Between 1919 and 1925 the population of rural East Prussia declined by 158,000, Pomerania and Silesia showing a similar trend, although not so pronounced.[1] Indeed the movement from the rural East Elbian areas to conurbations was so evident that one suburb of Hamburg became known as 'Big Mecklenburg'. The motive behind this was the quest for a new type of employment; of land workers who left Pomerania in 1929 only one in fourteen subsequently found agrarian work elsewhere.[2] Not all the shift was from east to west, since western holdings also suffered labour loss. On small farms in the Hunsrück/Eifel area of the Rhineland two-thirds of the labour force for milking consisted of help from outside the family in 1900, but by 1928 the proportion had fallen to one-third on average.[3] The Republic made some efforts to halt the continual drain of rural labour by instituting vocational training schools for landworkers to give them a better range of skills, etc. and the chance of acquiring a diploma. This does not seem to have met with any real degree of success, however, since between 1925 and 1933 the agricultural working force declined by 713,000 for Germany as a whole.[4]

However, this conceals that at the time of the NSDAP's accession mass unemployment in the cities was to some extent counteracting the previous trend, if only temporarily. Rural areas had only about 5 percent of the total workless in January 1933. In official circles the conviction prevailed that there was no prospect of a labour shortage in agriculture, and one minister stated that requirements could be met entirely by German nationals in the foreseeable future.[5] A contributory factor was the huge increase expected in the number of school-leavers in the near future, as the children produced in the postwar 'bulge' were due to leave school in 1934.[6]

It was decided, therefore, to try and kill two birds with one stone by installing the unemployed, especially youth, as rural helpers to assist the over-worked farmer's wife, and relieve the pressure on urban jobs simultaneously. Consequently via the Hitler Youth a scheme was initiated, the age-limits for the help engaged being from fourteen to twenty-five. The farmer was not obliged to accept the labour offered, nor was the worker compelled to take the job. The programme came into force on 1 March 1933, and was based on the payment of board and social security contributions plus wages by the farmer, in return for which he received RM 25 monthly from the state as a subsidy. The party press enthused over the project and how it was reversing the flight from the land. One case was quoted of a youthful helper from Berlin who had gone on farm aid to the very place from where his own father had migrated to Berlin thirty-two years previously.[7] But none the less it is clear that the programme was not wholly successful; one farm journal ran a special article calling on local farmers to utilise the scheme, as only 20,000 in the whole of Westphalia had done so. By the following October peasants were being appealed to in this region to keep the helpers on in winter and so combat unemployment in the cities.[8]

In East Prussia under the administration of Erich Koch an almost farcical situation developed due to his enforced solution of the local unemployment problem at the expense of the peasants. At Osterode the regional administration called together representatives from the local farmers union, the Agrarian Office of the NSDAP, the party itself and parish chairmen, and simply assigned so many unemployed to each district, including a proportion to agriculture. Columns of men were organised to take part in land improvement and allotted to different peasants who had to feed them. No preparations had been made so that the participants did not know what to do. In Wehlau the *Kreisleiter* said to dismayed peasants in the presence of the local KBF that if they revealed what was happening they would be arrested. Another told farmers that if anyone refused to accept the unemployed they would be driven on foot to local party headquarters bearing a placard 'Saboteur of Work'.[9] Thus Koch reduced the unemployment statistics in his own fashion.

It seems unlikely that the situation in general was so curious as in East Prussia, and although the peasants were reluctant to take on unskilled labour, as many reports testify, no doubt some contribution was made by the scheme to solving unemployment and assisting in land improvement: in January 1935 as many as 11,847 emergency workers were engaged in land betterment, apart from individual helpers on farms.[10]

By 1934, however, it was clear that rural migration was becoming a fact of life again. So an Act in May empowered Labour Exchanges to withhold permission for a change of job to anyone employed in agriculture, provided that he had worked on the land for three years prior to the measure. If any such person had obtained new work without its consent, then his new employer could be compelled to surrender him. To justify this legislation it was pointed out that the proportion of unemployed in total accounted for by those out of work in the major cities was actually rising.[11] Hence the threat to return migratory landworkers even if they were lucky enough to find work in the cities: the unspoken implication is that their employment there would have deprived urban unemployed of job opportunities. The Act was an authorisation only, not an order.

It was nevertheless used on occasion, as in the town of Burg in autumn 1934, when it was announced that in the coming winter all resignations by landworkers would be scrutinised by the local Labour Exchange in conjunction with the KBF with a view to retaining workers on the land and so avoiding the need for foreign labour. Apparently 15,000 workers were returned compulsorily to the land between April 1935 and March of the following year after they were discovered to have taken other jobs.[12]

To complement this legislation another ordinance in May 1934 totally forbade the employment of landworkers in certain named industries (including metalwork and building) without prior Labour Exchange consent. Despite the return of some workers, however, the two measures seem to have failed to stem migration in general, even though the so-called 'Labour Book' was introduced in 1935 for all workers. Apparently this measure also was at least partially determined by the need to stop continuing rural migration.[13] This legislation, however, had little effect in practice. By late 1935 landworkers were said in one part of north Germany to be deserting their posts in order to get jobs as machinists, etc.; the possibility of a labour shortage in the area was already manifest. Almost coincidentally the Minister of the Interior in Bavaria authorised the recourse to concentration camps for farm-workers who broke their contracts.[14]

Of course, the existing regulations were sometimes applied, and many former landworkers were compelled to give up a lucrative job outside agriculture and return once more to the fold. The KBF at Niederberg (Rhineland) was very active in his attempts to get back the runaways, as witness his letter to a local farmer asking him which construction company a former landworker was employed in. On another occasion he wrote to the Labour Exchange in Düsseldorf to get back a man now employed in a brickworks.[15] The whole procedure sounds like the lord of the medieval manor trying to trace his runaway serfs; *Stadtluft macht frei* (the city makes you free) seemed to be repeating itself as a slogan for the agrarian labour force in the Third Reich, as it had once done for its counterpart in the Middle Ages.

Despite the energy displayed by some KBFs by 1936 it was apparent that the laws were actually turning out to be counter-productive, as no one would take up agricultural work as a career for fear of being caught there for good.[16] Parents in rural areas were sending children to the cities as soon as they left school. The measures were therefore suspended that year; they can only really be described as having had a negative effect by having frightened so many away from the land as a future career. Ironically this was apparent even to the busy KBF at Niederberg, as he pointed out to his superiors at the time.

As the work-creation programme, and rearmament, got further under way, rural migration began to intensify. As early as February 1935 the press was acknowledging it as a fact of economic life: three years later the trend had so accelerated that the RNS produced its own survey in respect of farms up to 50 Ha in size. The result showed conclusively to what extent landworkers' and peasants' children had been voting with their feet in a vocational sense in the Third Reich. If the agrarian labour force index of such holdings was 100 in 1935, three years later it was actually down to 70, a catastrophic fall for such a relatively short time-span.[17] Even more striking was the uniformity of the movement. Only one of nineteen *Landesbauernschaften* examined still retained 85 percent of its 1935 labour force; this was in isolated East Prussia, where the military build-up was on a somewhat smaller scale, and where above all there was no autobahn programme. The conclusion is obvious; only where no other job opportunities existed did people remain on the land. This impression is reinforced by statistics from areas where industry and agriculture stood in close proximity. For example, in the Cologne district holdings under intense cultivation, i.e. dairy farming, root crops and horticultural enterprises, had less than half the outside labour available in January 1938 than three years previously.[18]

By then the whole question had become easily the most important single problem confronting Darré, as shown in his speech to the National Peasants Assembly. He estimated that the total of rural migration, comprising owners, family helpers and farm labourers, amounted to between 700,000 and 800,000. He was supported by the census of May 1939, which gave a figure of nearly one and a half million, the overwhelming majority of which was male.[19] These statistics, however, related to the rural population as a whole, so that the actual labour loss was considerably smaller. It was none the less serious, and an obvious blow to the *Erzeugungsschlacht,* especially as the total numbers employed in the economy as a whole went up by some two million between 1933 and 1939. In other words, there was rural migration when the NSDAP came to power and it continued at an increased rate during their first six years in office. These figures are naturally generalisations, in that not all districts suffered equally, as their statistics are concealed within the *Landesbauernschaften* totals. In some cases there was actually more labour available in 1939 in an individual district than previously, but they were exceptions to the general rule.[20]

Given that the NSDAP had always made a special fuss of agriculture, how did it come about that it failed to arrest the decline, at a time when every particle of food was vital to the country? As rural migration was clearly such a large-scale phenomenon, a mono-causal explanation would almost certainly be inadequate. In fact, it would appear that low wages, poor accommodation, lack of modern amenities and general overwork on the land were all contributory factors.

The role which bad housing might play was recognised early by Darré, and in January 1935 he wrote to the Finance Ministry pointing out that if the *Erzeugungsschlacht* was to be successful the necessary labour force must be procured; since accommodation was a problem he asked for a subvention of 40 percent for new building costs. He did not get far, and took the matter up again some months later, alleging that

18,000 married landworkers were unemployed because they could not find quarters.[21] His promptings had apparently little effect and nothing was done on any real scale until the arrival of the Four Year Plan. Goering then announced loans at 4 percent to aid rural housing; the amounts advanced varied with the farmworkers' status. The problem still remained intractable, however, as the Third Reich simply had other priorities. From 1933 to 1938 only 23,338 new dwellings were provided for landworkers, a mere drop in the ocean. In the following year it was calculated that as many as 350,000 were still required.[22] All contemporary sources stress the poor living-conditions for the rural labour force and the part which they played in intensifying rural migration.

Again, the NSDAP was trying to combat decades of inherited neglect, since it could hardly be held responsible for a situation prevailing when it took office. As has been seen (in Chapter 6) both the RNS and the DAF did what they could to improve things, by bringing pressure to bear upon employers. But what was really required was a large-scale building drive, for which neither the finance nor the labour was available once rearmament got under way. Once the decision to build up the armed forces was taken as the first priority agriculture had to yield its precedence.

Of course, the farmers themselves, and especially their wives, suffered from the poor conditions too. Nearly two-thirds of farms had no running water in the thirties. Consequently the time spent just at the pump could amount to 700-900 hours yearly, at least in Saxony, and presumably elsewhere. This was the core of the problem; rural life was simply no longer attractive because primitive conditions inevitably meant overwork. A sample of ten holdings in Württemberg revealed that the occupants were working 3,554 hours per annum on average – even more for the unfortunate wives. Hence the rhetorical question posed by one contemporary observer: 'How many sturdy, healthy, blooming peasant-daughters become, only a few years after marriage, haggard exhausted old women?'[23] The general effects of overwork on their health was considerable. Behrens, writing of Saxony, mentioned reports of horrifying rural conditions piling up on his desk, and gave two examples of overworked wives suffering miscarriages.[24] Another source actually quoted a village where in one year over a quarter of pregnancies terminated with a miscarriage.[25] Faced with such a future, women were ready to leave the land even for lower wages, as when a paper factory opened in Pomerania. In Gunzenhausen (Bavaria) a paint-brush concern took away female labour although its wage-rates were not high.[26] For those who did stay by virtue of having already married peasants, it was evidently more a question of resigned acceptance of fate than of looking forward to any real enjoyment of life. So strong was aversion to farmwork that by November 1938 at the Annual Peasants Assembly, Darré was launching a desperate appeal to farmers' daughters not to migrate and leave their mothers in the lurch.

The RNS, the *NS Frauenschaft* and even the Labour Corps did attempt to lighten the workload for farmers wives, by giving advice on household management, farm chores, apprentice training, etc. for which Section IC and IIH of the RNS were responsible. The Labour Corps had a department especially to help settlers' wives to accommodate themselves to their new homestead. Similarly the party welfare organisation (NSV) arranged for creches and kindergartens to be installed in villages in agreement with the RNS, the object being to enable the women to leave their children under supervision whilst they helped their husbands with the farm. The NSV itself undertood the task of appointing a child supervisor, whom the RNS introduced into village life by arranging contacts at a social level between her and the local people. The Reich government made funds available for the actual kindergartens.[27] The scheme in principle was excellent but the degree of overwork for women was such that it could never have sufficed in itself. In any case, the work of the party organisations was in itself counter-productive in that as well as help and advice, ideological training also formed part of the programmes (of which more later). From the standpoint of the farmer's wife this was time-consuming and counteracted the useful assistance offered in other areas.

Since all hands were needed to get the utmost possible from the soil, even peasant children suffered from overwork, and its effect on health. Three-quarters of all rural children inspected in current surveys had deformed backbones, and nine-tenths had foot defects of some kind or another; in one country school in Saxony there were only seven children completely fit out of ninety. This had serious implications for any military programme: the March 1930 programme had called the peasant the backbone of national defence, but the reintroduction of conscription showed just how deficient rural youth was in physical terms. Examinations of recruits showed that those from the cities were superior to their country cousins, which was ascribed to poor hygiene and overwork for the latter.[28]

It is not surprising that German youth did not actually rush to take up farming as a vocation. This was the worst aspect of rural migration from the national standpoint, since unless checked it nullified the entire 'peasant policy' including settlement. The loss was all the more serious as being qualitative as well as quantitative; it was the best of rural youth that left the land, as one report in May 1939 emphasised by stating that the well-endowed pupils who remained in agricultural schools were the exception to the rule. The LBF in Kurmark told Darré that the pick of the fifteen to eighteen-year-old rural youth *never* (underlined in the original) chose the career of a landworker. Wages and the gradual creation of more urban jobs were often cited as the greatest factors in deterring youth from the land, but some sources spoke of long hours and no free Sundays as being equally decisive in their effect upon both parents and school-leavers. Of 1,300 youths leaving school in rural Lippe in Easter 1937 only sixteen chose the land as a career.[29] Clearly an agrarian life had lost its attraction.

This was reinforced by the reluctance of girls to marry into the farming community, which in its turn induced the migration of youths. There was solid contemporary support for their belief that girls preferred a husband from another walk of life. An analysis of matrimonial advertisements in rural south-west Germany in 1935-36 showed 224 requests from peasants seeking a wife, which drew a like number of replies. But a group of forty manual workers, craftsmen and officials got 400 answers to their insertions in the same journal. The relative unattractiveness of farmers as husbands was underlined even more sharply the following year, when 71 percent of all prospective peasant bridegrooms found their requests unanswered.[30] From the other end of the country a similar state of affairs was reported by an OBF in Lower Saxony where a quarter of the unmarried farmers in his village could not find wives for the identical reason.[31] As an occupation agriculture apparently had no sort of status in the eyes of contemporary German girls.

In order to try and improve its prospects for young men the government encouraged special vocational training, a three-year course being terminated by the award of a landworker diploma. Instruction was in the hands of teachers who had once been farmers, and was directed by the KBF. There were also apprenticeships for specialised tradesmen, cowmen, shepherds, etc. part of the instruction being given on a designated farm, and part in a vocational school. No great enthusiasm seems to have existed for this, as of 41,000 such vacancies in 1937 only 7,000 were actually taken up by aspiring landworkers.[32] The scheme was admirable enough in principle but the lack of enthusiasm is surely significant.

Whatever rosy propaganda pictures were painted in the official press about life on the land, inner circles knew perfectly well what sort of resistance to an agrarian career existed among young people. By 1938 the idea of legal compulsion had been abandoned, judging from an official view which acknowledged the facts of the case and asked how peasant youth could be made to stay voluntarily on the land, underlining 'voluntarily'.[33] There did not unfortunately seem to be any answer to this judging from the masses of evidence in the form of Trustee of Labour, Gauleiter and local government reports from 1937 onwards. These often stressed flight from the land among young people as a simple fact of life.

The qualitative aspect here was worse than just its purely numerical side. By May 1938 in three Württemberg parishes two-thirds of all peasants were over fifty years of age. Of the three peasants' sons confirmed at Uelzen (Lower Saxony) in April that year, the only one intending to follow in his father's footsteps was described by the local *Kreisbauernschaft* as 'ripe for sterilisation',[34] an allusion to the 1933 law permitting compulsory sterilisation in cases where reproduction by the subject was biologically undesirable. This was by no means an isolated example, and boded ill for the whole future of agriculture. Moreover, many peasants' sons were refusing to take over the family holding at all. For example, from Bavaria it was stated that one *Bauer* had three sons, one of whom was in the army and had renounced the family farm, whilst the others had both said: 'I'd rather be a worker than take over the holding.'[35]

Some efforts were made to remedy the more glaring defects of village social life, quite apart from the construction of new dwellings. It was accepted that a brighter community might counteract the tendency to leave the land, and various organisations of the party co-operated in the production of leisure-time functions, including the Hitler Youth and the League of German Girls, but more especially the DAF. The rural population were said to have been grateful for this variety 'in their otherwise monotonous life'.[36] The last phrase goes a long way to explaining migration in an era where the cities with all their attractions offered a glamour which the countryside could no longer match. That the proximity of Berlin had more than just an economic influence upon landflight in the east was stressed by the LBF in the Kurmark.[37] The DAF set out to counter this with an organised campaign, 'Beauty of the Village'. Any place could qualify as a model village if it possessed a swimming-pool, sports ground, a Hitler Youth hostel, village hall and first-aid station. How many did possess all these facilities is evident from current statistics, since by 1937 fewer than 250 villages had qualified for the title.[38]

In the same year the DAF took over the main responsibility for the cultural entertainments in agreement with Darré, announcing that it would fight rural migration by giving the village its soul back. A campaign was launched by special staff who laid on evening shows, working with the LBF concerned as far as possible. The programmes appear to have been based mainly on poetry reading, theatre groups, marionette shows and music, both military and otherwise. The *Kraft durch Freude* movement (of the DAF) also organised sport as far as possible, which it started on the land in 1934. In the east this apparently met with little response among the rural population;[39] anyone who had to spend two hours a day or more just pumping water, apart from other tasks, was not likely to have much time or energy left for physical recreation afterwards.

Obviously there is nothing wrong in principle with attempts to brighten up the villages and provide them with decent amenities. But statistics indicate that the efforts were on a small-scale, had little effect, and however admirable in themselves simply could not compensate for overwork and poor financial rewards. What other sectors of the economy really offered was more money for less work, especially as the rearmament boom developed and job opportunities outside agriculture multiplied.

As the farmers' financial position improved and labour got short, agrarian wages did go up. In 1938-39 hired help received RM 1,846 million in cash wages as opposed to the RM 1,332 million six years before. But rises were limited by the needs of the *Erzeugungsschlacht*, as the peasants were expected to invest substantial sums in the holding. In 1937-38 their expenses amounted to RM 6.9 billion compared with RM 5.9 billion five years before, but most of the increase was in capital improvements, such as machinery and buildings or augmented use of fertiliser, as tax and interest payments had been cut. To put this in another way, the price-index for such items as tools, wagons, machinery and fertiliser was marginally lower in 1939 than it had been in 1936. The peasantry's investment in real terms had gone up sharply in obedience to government demands for greater production. The net result was that expenses were two-thirds of sales in 1934-35 but 72 percent three years later.[40] By February 1938 Backe publicly admitted that the point had been reached in the *Erzeugungsschlacht* where financial gains and losses were becoming equal for the peasants. So much new investment was needed that the profit margins were simply getting unsatisfactory, in other words. Clearly this inhibited further rises for hired labour, and encouraged it to leave for other, better paid, sectors.

One fairly clear solution would have been higher prices for farm products. What reinforced the Führer's refusal to grant any is a long story, beginning in 1934-35 when retail food prices went up anyway because of shortages. By summer 1935 the position was such that Goerdeler told von Rohr that he did not think that Darré would last out the year in office.[41] The pressure on him was indeed considerable. In July the Mayor of Stuttgart was complaining to Frick about food prices, particularly of meat, which especially perturbed him as he had so many similar reports from other parts of the country, notably Kassel and Wiesbaden. In the first the Gestapo stated that demands for higher wages were the result. Camburg (Thuringia) had apparently seen an increase in cattle prices of one-third in the last six weeks. From Kassel a 'certain unrest' among the workers was mentioned. For the country as a whole food prices were said to be 8.1 percent higher in August 1935 than in the same month of 1933.[42] This was rather more serious than it sounds, because this meant, in effect, a 4 percent cost-of-living upswing for manual workers; demands for a similar wage augmentation could have added to export costs and depreciated the value of the government's internal work-creation plans.

What strengthened its determination to pursue in effect an incomes policy was the tendency for wages to rise anyway in 1935. Goerdeler was among the first to draw attention to the inherent dangers of this for the whole programme of economic reconstruction and the Ministries of the Interior and of Labour were fully in agreement with his views. This was at a time when export prospects were officially held to be gloomy, in view of the international situation, so that internal inflation would have been the last straw. Indeed, there was even serious debate as to whether building wages should be cut, but a Gauleiters' meeting turned the idea down, as did Hitler, no doubt through fear of political repercussions among the workers.[43]

A further complication now appeared in the shape of the party, especially the DAF. Burgeoning industrial recovery was taken in an article in *Der Angriff* as the excuse to demand higher wages, let alone a diminution. This onslaught caused some agitation in governing circles and the result was a hurried conference on 2 May, over which Hess presided. Seldte pointed out that he had been trying to keep to 1933 wage levels wherever possible, but the DAF was now making it hard: rises in wages for specialists in the construction industry were threatening economic recovery. There was substantial agreement among all those present that the delicate currency situation, rearmament and the need for exports, combined to imply wage stability. Goerdeler referred to the continued presence of two million unemployed as another factor. It is informative to note that Darré was present and concurred, as did Hitler ten days later. No doubt the *Wehrmacht* build-up played some part in his feelings (as it did for Schacht and Darré).[44]

It was now up to Darré, having voted for stability, to accept it, and its corollary, no upward price-swing for agrarian produce. In fact only three months after the conference the Secretary of State in the Ministry of Labour was actually demanding lower food prices from Backe, in order to maintain the workers' standard of living, due to current price increases engendered by shortages.[45] No doubt this gave people in the cities as well as in the Civil Service the impression that the farmers were making a killing out of scarcity. But in effect the emphasis on a wages policy had already sealed their financial fate. This was especially true as the demands for stability at the May conference came from almost all those present, not merely from the Führer alone. Once the rearmament programme had been accepted it seems likely that any Right-Wing Cabinet would have treated agriculture in the same fashion.

By 1938 the financial position of the farmers was markedly unfavourable, due to demands for continually-increased investment without corresponding price revisions for foodstuffs. The wages policy and price stability prevented these in effect, as RNS leaders were aware, not that they necessarily approved. Reischle's pen on the subject became such an embarrassment as far as the Price Commissioner for the Four Year Plan was concerned that he tried to get some of his publications forbidden.[46] This did not deter the RNS leaders from pressing their claims for better treatment of agriculture. At the Peasants Assembly in October 1938 one spokesman pointed out quite frankly that the monetary expenses for the farmers were growing ever more disproportionate to their returns; he said openly that there was no real autonomy for the *Verbände* in the field of price-fixing, which was in reality politically determined. A more concise and accurate picture of the facts could hardly have been imagined, and the bluntness of the speech surely underlined RNS disillusion with the situation.

Behind the scenes pressure grew on the political leadership to grant agriculture a better deal. Darré himself pleaded for higher prices to Hess, who replied in part: 'I shall turn down all price increases until I get an order from the Führer to agree with you in this respect.' Rebuffed at the lower levels, the RBF now went right to the top with a letter for Hitler, in which he listed the aid which he felt the peasantry needed at that time. The total amounted to between RM 700 and RM 800 million annually, based on better prices for milk, beef, pork and flax, higher wages for landworkers and cuts in the cost of fuel and energy to rural consumers. Darré wanted the Finance Minister to release funds to soften the blow to consumers, in other words subsidised food. There appears to be no particular reason why this could not have been done had the political will been present; in 1938 Germany's total GNP was RM 102,000 million. Darré's suggestion therefore represented a relatively small sum. Hitler's answer was not long delayed; a minute in the Chancellery files recorded simply that: 'The Führer has taken cognisance. He is in principle opposed to all price increases for agriculture.'[47] Despite even this unequivocal statement of policy Darré did not entirely give up hope that perhaps a better deal could be obtained by further pressure; in January 1939 he despatched yet another memorandum to his leader, saying that the 'harmonious balance' in the economy had been lost; 'the cause of this is the undervaluation [financially] of agricultural work and its last consequence is the land-flight'.[48] This also achieved no result whatever, despite the fact that the last clause accurately summarised the prevailing position.

By now no one could seriously deny the consequences of the current economic miracle for agriculture. In Schleswig-Holstein demands were made for an extra RM 1,000 million yearly for dairy-farmers and stock-breeders.[49] The farmers' share of national income was falling steadily; from 8.2 percent in 1932 the NSDAP had raised it to as high as 9.7 percent three years later, but by 1938 it had dropped back to 7.8

percent, even lower than when the party had taken office.[50]

A vicious circle had been created whereby investment needs and government-dictated price stability held down the agrarian index. Farmers could not match wages in the construction industry, so both peasants' sons and farmworkers left the land. One current estimate attributed half the growth of 800,000 in the building trade's work force between 1932 and 1938 as being due to migration from agriculture. Consequently new decrees forbade the employers in the construction industry to engage workers from elsewhere without written permission.[51] Like previous legislation this seems to have had little effect; the NSDAP found to its cost the difficulty of trying to reverse the operations of economic forces by legal means.

The consequent drain of labour clearly worked in a way detrimental to the *Erzeugungsschlacht;* there is massive documentary evidence testifying to this for the period immediately prior to the outbreak of the war. The constant battle to keep up output targets in the context of a shrinking labour force reduced peasant morale so seriously that peasants felt themselves to be 'the beasts of burden for society in general'. One OBF was quoted as saying in 1939 that the hardest thing which he and his colleagues had to combat was their inferiority complex.[52] This was undoubtedly induced by the poor conditions in agriculture in general as compared to other sectors of the economy. Above all, they were deeply conscious of the low priority which they received, exactly the complaint always made under the Republic. As the *Landesbauernschaft* in Bavaria said, no one could close his eyes to the speed with which projects were carried through for other sectors compared to the 'snail-like pace' accorded to those for agrarian development.[53]

Owners in such a situation could either sell or let the property, as was frequently done, or reduce their stocks or their output in some way. That the latter course was often chosen can be seen from Darré's speech at the sixth Peasant Assembly in 1938 when he attributed an actual fall in butter production principally to labour shortages on peasant holdings. The deputy RBF could quote instances of stock-reduction in Saxony, where one farmer cut his cattle holding from ninety-five to sixty-five, whilst another could maintain only three-quarters the number previously owned. Even if stock was not reduced no advance in production was possible, despite the needs of the Four Year Plan; by January 1938 the deputy LBF in Bavaria was telling his superiors quite frankly, 'One cannot think any longer about increasing yields.'[54]

This appeared to be true all over the Reich, as current Trustee of Labour reports make clear. Of Thuringia, Hessen and the south west it was stated that in February 1938 the spring sowing was already endangered by labour shortages. In the Rhineland only old people and children were visible in the fields: farms which had had eighteen cows in the previous year now had two, and kept pigs for domestic consumption only. By the fourth quarter of 1938 instances were being quoted of holdings that required ten to fifteen workers having only two or three only, so that land under the plough since 1933 was being left fallow. What that meant for the *Erzeugungsschlacht* is clear enough. In January 1939 the situation in Lower Saxony was such that peasants openly discussed going over to less intensive cultivation. At Burgwedel one *Bauer* with a 64 Ha farm was left alone with a wife and two children under ten years old: of his three helpers, one was in the Labour Corps, one in the *Wehrmacht* and the third had given notice. In Kreis Celle peasants had refrained from reporting corn thefts by their own workers to the police for fear of losing the men altogether.[55]

If higher food prices were rejected and material means in the shape of attempts to improve housing and amenities had failed, then there seemed to be only two choices left. Either a drop in production was accepted, which would have jeopardised the whole economy, or labour must be obtained from elsewhere. Efforts were made in the years prior to the war to find replacments, of which the biggest single group was that of the foreign workers. There can be no doubt that this was a bitter pill for any nationalist party to swallow, especially as the 1930 programme had proclaimed that by bettering wages and living standards for landworkers and elevating agriculture in general and 'by a prohibition on rural migration foreign workers will be unnecessary and therefore forbidden in future'. The party initially observed this, as in May 1933 it was announced that no seasonal workers from abroad would be given permits until further notice. In the following year it was made clear that one reason for encouraging the Hitler Youth to take up work on the land was the desire to cut out foreigners.[56]

By 1935, however, simple practical necessity had already begun to triumph over ideology, since from 1 April of that year to 31 March 1936 over 50,000 permits for foreign labour for agriculture were issued.[57] The shortfall in domestic labour supply then deteriorated further and amounted to 374,000 by autumn 1937.[58] The Ministry of Labour then found it necessary to issue a special directive regarding the treatment of foreign workers, who were now becoming once more a regular feature of the German agrarian scene. In that year 95 percent of the temporary harvest labour in the Erfurt area came from abroad.[59] By the beginning of the next year the situation had worsened yet again from a nationalist standpoint. It was now calculated that the country would need some 200,000 foreigners to fill the gap for that year; ultimately only 80,000 were taken on because of currency difficulties. This was a point which did not escape the authorities at the time. Between devoting financial reserves to hiring foreign workers and using them for imports of food there was in the final analysis little real difference. No doubt it was mainly for this reason that the Labour Commissioner for the Four Year Plan described foreign helpers as being no real solution.[60]

By 1939 the influx had almost assumed the proportion of a veritable invasion with Italian workers even wanting to bring their families. Other nationalities involved included Dutch, Yugoslavs, Bulgarians, Poles and Hungarians. Additional to the officially-permitted stream of immigrants there was also a clandestine trickle. As early as 1937 it was estimated that at least 1,500 Poles had entered east Prussia in this fashion, with a similar number in Silesia. Two years after that the unexpected non-arrival of the official Polish contingent caused severe problems, so necessary had immigrant assistance become by this time. As a result farmers were asked to re-distribute workers of other nationalities among themselves to ensure fair play.[61]

All this was highly unpleasant from an ideological standpoint, bearing in mind the sudden arrival of foreigners, many young, in villages from which eligible German males had migrated. The consequence was intermarriage on some scale, at least in Lower Saxony. This has to be seen in conjunction with migration of the more gifted German youth leaving the less fit, even sterilised, behind. Agrarian labour was now becoming a mixture of these latter with foreigners, both classifiable as undesirables from a National Socialist viewpoint. Just what this implied for the country's future was put succinctly by one party official, 'The racial value of the rural population must decline.'[62] As by definition the latter was the life-source of the whole nation, this could hardly have been a heartening thought for the ideologically faithful.

Apart from external sources other reserves tapped for the land, particularly at harvest time, included the 'Duty Year' for girls, the Labour Corps, volunteers from various youth movements and even military conscripts and government white-collar workers. This labour was largely untrained and was at best a poor substitute for the skilled labourer who had migrated to the nearest building firm.

The 'Duty Year' for girls, applicable to all those unmarried and still less than 25 years old, was introduced in 1937, and had to be devoted either to agriculture or to domestic service; without a certificate confirming completion they could not obtain other employment. From 1 March 1939 until 31 January 1940 nearly 40 percent of all girls conscripted chose agriculture, which represented a powerful addition to the labour force numerically, if not in quality.[63]

A volunteer service also existed, organised by the League of German Girls (BDM). These helpers received initial training in groups, one per farm, to initiate them into the rural mentality, after which about 90 percent were ready for allocation on an individual basis. The life did not appeal to them and in Brandenburg only 6-8 percent remained after the first year's service. There were no legal limits to their working hours and some peasants were irresponsible enough to exploit the girls as cheap labour.[64] The masculine counterpart, the Hitler Youth, had its own Land-Help scheme, with volunteers from 14 to 25 years old, put only on peasant holdings as the larger estates used seasonal labour mainly in summer. By 1937 there were over 50,000 volunteers under this scheme, few remaining when their tour of duty expired.[65]

In addition to services for the individual sexes there was also a joint organisation called Land Service, based on mixed groups of ten to twenty in number: they lived communally but assisted on holdings as individual helpers for nine months commencing at Easter. The Land Service provided 26,016 workers in all in 1938. The volunteer services cannot be said to have made much difference at a time of such a chronic labour shortage, and the smallness of the number who did stay on is an indictment of rural conditions in general. Attempts were also made to interest children in the land by giving them nine months in agriculture when they left school; few took up the career permanently, despite this encouragement, as 'the parents' resistance was too strong'.[66]

Probably the largest single source of outside aid, apart from foreigners, was the Labour Corps; originally voluntary, it was made a compulsory service in February 1935 and its numerical strength became considerable. It was utilised mainly on land drainage, flood protection, reclamation, etc. A contemporary observer estimated its contribution to agriculture as being worth some RM 60 million annually. This conceals the fact that the Labour Corps was ultimately of little help to the man who needed him most, namely the peasant. This arose partially because of the quasi-military structure of the Corps and its unsuitability for the provision of individual helpers on small or medium holdings. Moreover, as the remuneration requested amounted to 40 Pfennigs per hour per worker, even the *Bauernführer* sometimes turned the Corps down as a labour source.[67] Efforts were made to overcome these problems by sending troops of helpers from village to village, which was unsatisfactory, since at harvest time all peasants wanted the assistance at the same time. In any case the Corps actually took labour away from the land by calling up peasants' sons, who had to serve in it as a result of an agreement in 1935 between Darré and its leader, Hierl. Even landworkers were conscripted, although in the Rhineland at least they were allowed to join on 1 October rather than 1 April like everyone else, which at least saved them for one summer.[68] In assessing propaganda about the Labour Corps and its projects it should be borne in mind therefore that behind these impressive-sounding statistics lay the reality of farmers battling with the problems of maintaining output at a time of labour shortage whilst their own sons were in the Corps and unable to help.

A desperate situation led to desperate remedies. To bring in the harvest any pair of hands was good enough. At Jülich the army was used in July 1937, and at Düren the shortage was so obvious that even in early summer a forecast was made that primary school-children would be required. The following year saw an appeal from Hess to the party and all its affiliated bodies such as the Hitler Youth, BDM, SA and SS to assist; apparently this was answered, as it was later reported how grateful farmers were for the help rendered. The same harvest time saw 1,500 sailors being employed on the land. In 1939 things were so difficult that white-collar workers in government service with some knowledge of farming or with relatives on the land, were granted two weeks special leave if they wished to gather in food supplies. In Lower Saxony one peasant just sentenced to a year in prison for attempted rape got three months suspension due to the labour shortage.[69]

The phenomenon of rural migration is a long-standing trend in modern countries, to which Germany prior to 1933 had been no exception. NSDAP ideology held that the peasantry was the life-source of the whole nation, so that in any case its duty was to supply population to the cities as well as replenishing that in rural areas. But what happened in the Third Reich went far beyond that and indeed exceeded any previous drain to such an extent that the whole *Erzeugungsschlacht* began to grind to a halt, as the deputy RBF conceded (see note 24). The landflight had clearly been in operation from the moment that unemployment in the cities began to diminish and accelerated once the Four Year Plan intensified the construction boom. Labour shortages subsequently lowered peasant morale and dashed the hopes of 1933 when it seemed that at last a government was in office which would put agriculture before exports and international trade. By 1938 it was clear that the peasantry was now playing second fiddle to defence instead. In Pfarrkirchen (Bavaria) morale was so low that peasants were saying that they could now see why the *Erbhof* law had been created, namely, to tie them to the soil and ensure that their farms could not be sold. It sounds suspiciously as though they would have done just that had they had the chance. The deputy LBF in the region stated candidly that there was only one solution to the process which left only the old and unfit on the land, and that was 'a slowing-down in the tempo of rearmament'. The worst effect upon those remaining was passivity in the face of the Four Year Plan tasks.[70] By February 1939 Bavarian peasants had become so desperate that they staged a public demonstration in Munich by marching from the Labour Exchange to RNS headquarters in the city.[71]

Apart from the effect on output, the other two aspects of rural migration worthy of comment are the attitude of youth towards the land, and the inability prior to 1938 to stem the landflight by legal means. In July 1938 Goering produced a measure allowing Labour Exchanges to send workers to certain jobs, but only for a certain period of time. The latter condition was removed in the following February, so that another step towards full control had been taken. The previous month the Four Year Plan's Labour Commissioner said quite frankly that 'absolute freedom of movement cannot exist in a totalitarian state'. Duty to the community was to rank higher. In March a further decree made Labour Exchanges' consent a necessary prerequisite to a worker's dismissal by his employer in particular industries, including the agrarian.[72] The full apparatus for the total direction of labour was gradually being constructed. As relatively mild legal sanctions had not worked, they would have to be made more severe.

Linked with these growing restrictions on vocational mobility was the youth question. As has already been emphasised, legislation in 1934-35 had proved not merely abortive but counter-productive. The harder it was made to leave the land the less likely young people were to embark on an agrarian career in the first place. So short-term restraints on job mobility militated against solutions in the more distant future. It is rather difficult not to feel that the NSDAP's dream of the peasantry as the permanent cornerstone of national life would have become progressively harder to realise with the course of time. Moreover, it was not only outside labour which was reluctant to stay in agriculture, peasants' sons and daughters were migrating too. Sometimes there was full parental encouragement for this, as when one peasant virtually ordered his son to remain in the army as a career, rather than take on the holding.[73] In face of such feelings any attempts at legislation or compulsion in any form seem all too reminiscent of King Canute and the tide. The facts of the economic situation blunted even the weapons against bad farming. When one *Bauer* was removed from the management of his holding for incompetence he got a job in construction at once, as a gang-leader at RM 350 per month, despite his lack of experience. He soon reappeared in the village pub in new clothes and with a style of behaviour commensurate with his new-found prosperity.[74] For bad farming to be the path to an easier life is hardly the best incentive for an efficient agriculture.

13

NATIONAL SOCIALISM AND PEASANT CULTURE

Race and Romanticism on the Land, and Political Education

In March 1930 the NSDAP had promised to pay due attention to peasant life in their agrarian programme, of which section 4 was headed 'The peasantry should be elevated economically and culturally'; its final paragraph spoke of the 'revitalising of peasant culture'. This was public recognition of the fact that there was more to any peasant policy than merely guaranteeing prices and reforming the inheritance laws. The farming community needed to acquire a new dignity and importance as well.

Quite soon after his appointment as Minister, Darré took steps to implement the new policy by nominating Erwin Metzner as a special official concerned with rural customs. What this implied was made clear when Metzner attacked liberalism for having pushed peasant culture into the background for economic reasons, e.g. cheap ready-made suits had come to replace traditional costumes. It was the task of the NSDAP to bring a fresh blossoming on the land in cultural terms. Similarly Motz of the 'Blood and Soil' Section of the REM spoke of his duty to free the peasant from his inferiority complex and make him conscious of his own worth. Meinberg rammed the approach home at a mass demonstration in Hamm; under liberalism the word *'Bauer'* had become an insult, the object of fun in Jewish comic papers. Now the peasantry would become the very foundation of the state: 'under Adolf Hitler we want to build up Germany as his peasant militia'.[1]

There is no reason to suppose that the landed population found all this sudden attention anything but welcome. After all, it went hand-in-hand with material assistance, which surely must have made it appear sincere. Additionally, every possible action was taken to safeguard the peasants' image in the Third Reich. The rural population was always presented in a favourable light to the rest of the community: plays or books that made the peasants seem ridiculous were violently attacked. In Trier for example, the aid of the president of the National Chamber of Authors was invoked to call in a book deemed to be offensive in this respect; the RNS official responsible asked subordinates to be on the lookout for similar cases.[2] A Cologne revue in 1935 was bitterly attacked for the way in which it portrayed the *Bauern.*[3] Every effort was made to ensure that the peasants received due recognition in the mass media in general.

Apart from acting as a kind of self-designated watchdog of the agrarian image the party was adept in the preparation of elaborate ceremonial occasions in honour of the farming community. The two most noteworthy were the Harvest Festival *(Erntedankfest)* and the annual National Peasants' Assembly *(Reichsbauerntag).* The first was instituted by a new law in 1933 naming the first Sunday after Michaelmas as Harvest Festival, a special occasion that had nothing to do with church practices and was organised by the Propaganda Ministry, not the RNS. At the first ever such assembly as many as 700,000 peasants were said to have attended to hear Hitler, Darré and Goebbels at the Bückeberg, near Hameln: as a preliminary a delegation of *Bauern* visited the Chancellery to present the Führer with samples of local food specialities and peasant art.[4] The whole nation was brought into the ceremony by means of the radio, and there were processions in various cities: in Koblenz, for example, Rosenberg spoke at the *Deutsches Eck.* Harvest Festival was repeated annually as a propaganda exercise on a large scale, partly in order to publicise the services of agriculture to the nation.

The annual Peasants Assembly, held much later in each year, was more of an internal event, run by the RNS. Its location after 1934 was Goslar, which was also the permanent home for part of the corporation's headquarters. This picturesque and historic town was chosen for its location, in Darré's own words, 'in the heart of the original German peasant land'. The transfer of the office to the 'core of the old German Reich of the Saxon emperors . . . in the vicinity of the cradle of the peasant duke, Henry the Lion' was said to symbolise that the peasants' fate would no longer be decided in the 'asphalt desert' of Berlin. It may well be that Darré had a more practical reason for the move; Goslar had given the NSDAP an absolute majority in the 1932 elections. The party's attachment to what ultimately became known as the National Peasant town *(Reichsbauernstadt)* was rather more than mere Romanticism; incidentally it made both Hitler and Darré freemen of the town.[5]

Unlike the Harvest Festival the annual Assembly was more than a mere propaganda exercise. In 1934, for example, three days were given over to talks on purely professional and technical aspects of the RNS organisational structure. However, Goslar itself did have some link with the Harvest Festival as well, in that delegations of *Bauern* who received special honours in view of their productivity in the *Erzeugungsschlacht* were presented there to the Führer at the former Imperial Palace in connection with the Festival: deserving landworkers were also introduced.[6] But the main point was always to give out the guide-lines of official policy objectives to the hierarchy of the RNS itself.

The main task of furthering peasant culture as such fell upon Metzner's new section, which existed at every RNS level except the lowest. Its duties were the awakening of appreciation for the peasants' way of life in general; the *Kreisbauernschaft* was to collect anything pertaining to rural folklore, including genealogical research. At the level of the *Landesbauernschaft* there was to be publicity for special cases and exhibitions would be held.[7]

The officials went to work with a will, and every aspect of traditional village life was investigated, for example, that of costume. A lady in Thuringia was commissioned in 1935 to produce suitable dress for the local peasants, traditional garb being taken as the model: the products were advertised in the national *Bauernzeitung.* Such customs had been better maintained in some regions than in others. The RNS in the Rhineland made inquiries of its counterpart in Kurhesse with a view to copying its local dress for Rhineland peasants. Within the latter area itself visits by costumed groups were arranged between villages to stimulate local interest.[8] Clearly such activity in the thirties in a relatively industrialised country meant that the RNS was walking on a rather narrow tightrope between the picturesque and the almost laughable. Even to the peasants some of the emphasis on medieval garb must have appeared old-fashioned, exactly because they had already abandoned it of their own free will, as some officials were well aware. As one put it, in describing attempts to 'dig out' old costume in districts where it no longer constituted a living tradition, the step from the sublime to the ridiculous was very small.[9] Indeed the peasants who wore local dress to the Harvest Festival often looked rather self-conscious when photographed in the streets of Hameln, judging from contemporary newspapers.

Another tradition marked out for attention was the local dance, as witness the essay 'On the old Germanic dances' in the RNS's own news service. Groups of these were presented at the Peasants Assembly; precise instructions were laid down as to their height, both for girls and boys and hair colouring, which should be blond. In other words, participants should look as Germanic as possible to heighten the effect. Membership of all such groups in the Rhineland was similarly based, in accordance with the racial criteria adopted by the SS.[10] The new settlers photographed in the east seemed to be of a like appearance, although this does not necessarily imply that Nordic looks were an essential prerequisite to getting a holding, merely that more publicity was given to persons of this type. Peasant weaving was another form of traditional activity to be fostered. Samples rated as locally typical were sent to a national RNS exhibition devoted to the theme.[11] The smiling peasant girl in traditional dress bending over a loom was a characteristic feature of RNS publications, to symbolise apparently the solid values of the whole way of life on the land as opposed to urbanisation. How many hours a week they worked and what they looked like at thirty was discreetly overlooked, at least in the more romantic of the news-sheets.

One final aspect of the campaign to revive folklore was the use of old Germanic names for the months, for example, *Julmond* instead of December. This meant nothing to contemporary Germans, so that even *Odal,* the RNS's own magazine, printed the modern name in brackets, as did internal correspondence in Darré's organisation. However, this idea was not original, as many völkisch groups had done the same, even before the NSDAP had been founded. What was new was the change of title of the journal itself from *Deutsche Agrarpolitik* to *Odal* in 1934. The latter word was ostensibly the name given to the old Germanic peasant holding, so that the alteration represented another piece of the kind of Romanticism so typical of the whole RNS approach.

But this kind of change does raise the question as to how genuine some of the old customs were that were now to be revived. Voices were not wanting in the Third Reich to suggest that so-called peasant lore was being taught on the land by townees, rather than being real rural traditions, which the RNS vehemently denied.[12] But there is no doubt that the party did invent new ceremonies and customs quite apart from the revival (and elaboration) of dead ones, as for instance the procedure at the hand-over of the family holding to the heir. There was no general custom in this case, so the LBF (Rhineland) then proceeded to give out general guide-lines for his area. This function was to be on quite a large scale, with the whole village participating, as well as the entire *Bauer* clan. The KBF was to make a speech on the history of the family and its farm and then hand the heir an appropriate symbol: the new young housewife was then to relight the hearth fire, using that flame brought to her from another nearby *Erbhof.* The KBF then solemnly reminded the new owner of his duty to clan and people: after this the legal conveyance followed; only peasant costume or uniform was to be worn.

Other parts of the country saw similar procedures carried out, for example, at Rodenbach (Thuringia) where a young farmer handed the heir a decorated spade as a symbol of the working German with a speech to the effect that the companion of his youth now became *Bauer* on the holding of his ancestors. 'May it [the spade] never rust and you never rest until you give the farm to your heir.' The father gave soil and water, and the house key to his son to symbolise what he had now acquired, i.e. fields, woods, house and brook. The KBF thanked the retiring owner for his services and reminded the newcomer of his responsibilities: here again a torch from a nearby holding was used to light the hearth-fire by the new housewife.[13] The whole affair was in sum virtually a religious ceremony in which the clan and people replaced a deity: the solemnity must have been quite impressive, and presumably again welcome to a sector of the community not used to receiving such tribute to its importance. There is no reason why thanks on behalf of the community should not be rendered where they are due, and the continuity of an agricultural calling stressed. Participants may well have felt themselves to be honoured in this way. Certainly the whole procedure shows the extent to which the RNS carried out its intention of trying to elevate the *Bauer's* standing in society, and with it his own self-confidence.

This comes out more plainly still in the case of another event, designed not just to pay homage to the peasant in his own immediate circle but also to attract the attention of the public at large. Those honoured were peasants who could prove occupation of the same holding for 200 years in their family line (150 years in the case of those in the frontier areas).[14] Procedure was unified for the whole Reich; the ceremony applied to both sexes, and the recipients were to get a certificate signed by the RBF and from the *Landesbauernschaft* an appropriate commemorative plaque in wood or enamel: such families were also honoured nationally at the Harvest Festival. For the actual local event a day-long programme was instigated; over 300 families from all over Württemberg were honoured in this way at Ingelfingen in 1936. There was a special play *Der Erbhof,* a procession through the streets and sports and a dance to finish, with an exhibition of old deeds, genealogical trees, etc. in the town hall.[15] To give full publicity to the occasion there was a radio report from Stuttgart the following day. The whole clan was included, incidentally, not merely the farm occupants, with a 'Clan Day' in the presence of the village, with the HJ, BDM and SA in attendance, the ceremony lasting thirty minutes, ending with a presentation by the KBF. Folk songs and dances formed a general background to the occasion.[16]

Large numbers of families were dignified in this way, as many as 1,500 applications having been received in 1939 for Württemberg alone. By that time the realisation had dawned at Goslar that the whole thing was getting out of hand, especially as it was estimated that there could easily be 150,000 families so qualified in Germany as a whole: it was suggested that either some less expensive method should be found, or that part of the costs should be borne by the peasants concerned.[17] Costly or not it seems very likely that the scheme was well-received on the land. One ceremony in Neustetten (Württemberg) so impressed a local peasant that he wrote to the *Landesbauernschaft* to record his gratitude for the event.[18] There is therefore no reason to suppose that it did not attain its objective of raising peasant morale.

Heraldry for rural dwellers formed another side of the cultural drive. Individual farming families were encouraged to submit a coat-of-arms to the RNS, presumably in line with Darré's own thoughts about a new aristocracy for Germany founded on a free peasantry. Some commercial firms apparently realised that an opportunity existed here for exploitation, and began to give out 'traditional' emblems, complete with certificates, on demand. By 1937 three-quarters of all those samples submitted by individual peasants in the Rhineland had turned out to be false.[19] There was apparently some quite genuine peasant heraldry, however, linked with crests on the family homestead. But it is difficult not to feel that this kind of activity was counter-productive. However much the RNS wanted to improve the farmer's standing, a coat-of-arms could only have made the owner of a 20 Ha holding look frankly pretentious. The commercial swindle merely highlighted this.

Any antiques or family heirlooms also served to build up the peasant cult and their presence at the homestead was eagerly catalogued. Similarly, *Bauern* were encouraged to give the holding a name if it lacked one, and the official farm title was solemnly entered in the Land Register.[20]

All this romantic enthusiasm was in many cases quite successful; the aim was summed up best by the description of one elaborate ceremony in the words: 'The *Bauer* again takes up his fitting place in the nation.'[21] The whole country was to understand his new significance; one local folklore commissioner sent a circular to all teachers in his area to orientate them in their task of teaching their children the meaning of rural culture. He did not wish to praise the past as such but to create a genuine contemporary peasant way of life.[22] The validity of this claim must be tested in the framework of costume revival, etc. What the RNS perhaps did not entirely face up to was the possibility that by about 1933 a 'contemporary' peasant culture simply was a contradiction in terms: the country had been industrialised too long. By 1895 there were already more people employed in industry than on the land. The enthusiasm of the RNS really came much too late in German history to be effective on a large scale, although many peasants doubtless enjoyed the fuss.

Quite separate from the question of raising morale on the land was that of political indoctrination. The RNS did not require that every peasant in Germany should officially be a National Socialist, as Darré himself pointed out.[23] But it did nevertheless attempt to win over the rural community for those aspects of its policy which it considered to be crucial for Germany. To this end courses had begun even before the formal accession to power. One such in Hessen-Nassau from 5-12 July 1931 had over one hundred participants, the curriculum including racial policy, autarky and the revision of land laws, with the usual attacks upon Marxism and liberalism. On reviewing it Darré suggested that social and technical matters should also be discussed; for the first, the peasants undergoing indoctrination in future should be afforded the chance to visit a mine or a foundry, and also a lunatic asylum or home for epileptics in order to bring home to them the need for a correct racial policy. Initiation into NSDAP ideology was thus well under way even before the Third Reich began; training courses in general became so successful that they were ordered for all Gaue.[24]

The drive to convert the farming community was intensified after 1933, with the objective of bringing the rural sector to a realisation of its 'German duty'.[25] By the latter phrase was meant not merely production increases but an acceptance of the whole ethos of National Socialism as such, especially the racial side. In other words, whether the peasant was to be formally committed to the party or not, he was to be so ideologically instructed that he complied with its policies anyway: between the concepts 'National Socialism' and 'German duty' there was in practice no difference except a purely semantic one.

At the Weimar meeting of the Agrarian Office in February 1935 Darré was at pains to remind all LGFs that the political indoctrination of the rural population was their task; in particular the *Bauer* must not be allowed to forget his völkisch duty and sink back into demanding 'interest politics', that is, those which suited agriculture alone (as he had done under Weimar). The general approach to peasant education has been described by one subsequent writer as resting upon the premiss that unless the *Bauer* was a National Socialist then he was no true *Bauer.*[26] This does not contradict Darré's opinion, already quoted, as it meant simply that for the party compliance with its views was more significant than actual membership.

The actual indoctrination was carried out mainly by the *Bauernführer* at local level, as they were in constant touch with the rural community. This aspect of their work was so important that the relatively large staff maintained by the RNS was justified exactly because political education was one of its duties.[27] The compulsory 'work evenings' for rural women were typical of this. A sample gathering began with a song, which was followed by some form of propaganda; for example, a lecture entitled 'Responsible choice of a marriage-partner', which was racial-biological in content. Two quotations from Hitler followed and then another song closed the evening.[28] Lectures given on these occasions were always based upon the NSDAP's view of life, and the 'work evening' is characteristic of the kind of incessant propaganda to which the agrarian population was subjected. (The NS Frauenschaft also ran ideological 'duty evenings'.)

Side by side with this went the constant pressure on the individual to conform, so that uniformity of both opinions and overt behaviour could be assured. One *Bauer* on the committee of a local co-operative in Coppenbrügge (Lower Saxony) failed to attend party functions and was hauled before the *Kreisleitung* to explain himself, where he expressed dissatisfaction with agrarian prices. The co-operative called on him to surrender his post, but the KBF eventually granted a year's probation.[29] Despite this, the drive towards conformity is obvious in the whole process. When one peasant tried to back out of a cattle co-operative he was promptly told not to play the outsider, otherwise measures would be taken against him.[30]

Proof of political reliability became a prerequisite of almost any facet of rural life. This applied to the controller of a *Verband* or to schoolmasters in agricultural institutions alike, as both had to produce evidence of their political views. Even an apprentice was expelled from an agricultural school because he took no interest in Winter Help or in the NSDAP in general.[31] The granting of a loan to landworkers to enable them to buy their own houses was dependent on proof of the right attitude, including the inevitable question on Aryan descent and details of the labourer's political background, such as information on any services rendered in the fight for 'national recovery', i.e. the NSDAP struggle before 1933.

The peasants' attitude towards Jews was carefully examined, which meant that they were expected to shun them totally in commercial transactions. In Lower Saxony the LBF in 1935 laid down that any peasant who dealt with Jews could not be given a prize for any RNS agricultural show; the money should be donated to Winter Help instead if he should win with his entry. However at Greiben (Rhineland) a Jewish ex-serviceman who had lost an arm in the First World War continued to enjoy *Bauer* patronage at his butcher's shop; none the less, the names of seven such customers were noted and sent in to the *Kreisbauernschaft*, presumably for future action.[32]

Although political conformity was imposed wherever possible, it does not follow that it was invariable; at least one *Kreisbauernschaft* declined to take political allegiance into account when honouring old-established families on the land. This was done precisely on the grounds that such discrimination would contravene the spirit of the whole operation, which should be to foster peasant interest in the history of the family and holding, with no other factors involved.[33] None the less, the final impression gained is that such acts were the exception, and heavy pressure on the peasants was the rule.

As a general policy this worked well enough in the short term, but the NSDAP clearly hoped that the future would be settled by indoctrination of the rising generation, in order that it was brought up from the beginning with the right outlook. This point was especially significant as the NSDAP had always had an appeal for youth. It was decided to strike this particular iron while it was hot, and guide-lines for co-operation between agricultural high schools and teacher-training colleges were given out by Bernard Rust, then Minister of Culture in Prussia.[34] The especial goal of the former was now to bring up the rural population to be conscious, responsible carriers of German society renewing its strength from blood and soil. Courses were therefore to include genealogy and eugenics, and instruction in the importance of heredity so that both physical and mental health could be maintained on the land. The frontier schools had a particular task in this respect: the will to settlement was to be encouraged in general in order to combat rural migration.

The policy statement for Prussia was soon followed by one for the Reich, with rather more detailed study plans, based on a series of subject groupings, of which the first 'People and Race' was more or less the same as that described above.[35] Then 'People and State' explained how National Socialism had changed the administration of Germany, including details of the organisation of the RNS. 'People and Society' dealt with the overcoming of class-consciousness and hence of the class-struggle in the Third Reich, and included details of the Labour Service. Then came 'People and Custom' which comprised folklore and associated themes, whilst 'People, Soil and Home' *(Heimat)* was a study-group based on the right of a nation to living-space, the history of German agriculture and of the country's colonisation of the east in the twelfth and thirteenth centuries. All this constituted a strong ideological slant in what was after all a vocational training school; education for the NSDAP clearly contained the concept of fitting the trainee for society as a whole, and by the very nature of the instruction given, ensuring that he would find such a society as the Third Reich congenial. The Germany of the future was to be populated by people conditioned to accept its ethos by the whole of their education.

So far was the emphasis upon indoctrination carried that in 1939 a textbook issued for agricultural trainees contained thirty-nine pages of political content compared to fifty-three devoted to professional training. One enthusiast even went so far as to declare that the signpost in the agricultural schools no longer carried the sign 'This way to rationalisation and profitability', but rather, 'This way to the peasantry and national culture'[36] *(Volkstum).* This was an exaggeration as the above textbook did after all still have more than half its pages devoted to technical matters, but there was none the less a solid core of truth to the assertion. Even the passing-out examination for apprentices demanded some ideological knowledge, as all candidates, after displaying their technical proficiency, had to give a talk on a political theme.[37]

Equally the official interpretation of politics penetrated into the village school. The kind of future education envisaged is clear from the results of a conference on the matter held jointly by the RNS and the NS League of Teachers in 1939:[38] it was suggested that there should be three broad courses based on the themes of blood, living-space, and what was in effect the history of the German peasantry. The first course included racial hygiene and genealogical research, plus the standing of the *Bauer* in the community: the second centred around the *Erzeugungsschlacht,* the Four Year Plan and the topic 'People without space' (the Germans). The last course was very much history as seen by the NSDAP, being concerned with the supposed link between the *Erbhof* and old Germanic farming, the colonisation of eastern Europe in early German history, 'the Jews in our village' and the thousand-year struggle for the Rhine. The emphasis on the earlier *Drang nach Osten* has interesting implications for any views on Hitler's foreign policy, as does the topic of 'People without space', a theme to which further reference will be made in the Conclusion.

Of all the various strands in the NSDAP's propaganda web none was more controversial than the repeated stress laid on race and breeding. 'Blood is the mysterious carrier of all qualities and characteristics of a race, of a species' ran the opening to one typical article.[39] Darré himself felt that race was the distinction between the Fascist outlook and that of the NSDAP; the Agrarian Office's task was to awake both the party and the peasantry to this. The Third Reich was not in his view a state seeking enhanced national production as its only goal like Fascism or Communism, it was a racial state first and foremost. With approval Darré quoted Hess's dictum from the 1933 Nuremberg Party Meeting that National Socialism was 'applied racial knowledge'; because of this Germany needed the peasantry; not economic motives but blood had led the NSDAP to unite it.[40]

The outcome of this approach was a vigorous supervision of the peasants' health in all respects, since for a healthy peasantry strict racial breeding was necessary, the latter being apparently for Darré the object of matrimony in the words, 'When one unites man and wife in the goal of child production . . . that's nothing more than breeding'[41] *(Zucht).* If the peasants were to produce healthy offspring for the country, then they must under all circumstances be fit. As one RNS organ put it, the race could only remain healthy when as strict a breeding was applied as in the case of plants and animals.[42] Medical records were thus compiled of the whole of the *Bauer's* family background, including such details as the cause of death of the grandparents.

The whole *Erbhof* family was physically examined (including from an anthropological standpoint) and any evidence of hereditary disease carefully noted.[43] Even spiritual and mental traits were recorded and blood samples taken. This procedure was for the *Bauer* only as other landowners, lessees or agricultural labourers did not undergo the rigorous check. The object of the blood samples was to determine whether the *Bauer* would produce healthy offspring. Additional evidence for the health of the entire family line was provided by the compilation of a Table of Ancestors for each entailed holding (a proviso in the EHG had foreseen this); the RNS 'Blood Questions' section was aided in the research by the NS League of Teachers and the party's Office of Racial Politics. Each *Erbhof* also had to have its own history book, compiled by the owner, relating to the whole family line, to bring home the importance of blood. All such documents had to be submitted to the relevant *Landesbauernschaft* by 31 March 1938. The drawing-up of a family tree assumed ominous overtones when it was agreed in the Rhineland to place these at the disposal of the Institute of Heredity, which already had records of illnesses of over 500,000 people extending back over one hundred years.[44]

For the policy of eugenics, which is what all this campaign amounted to in effect, has two sides, one of which is the furtherance of that held to be desirable, in this case the healthy peasant. The reverse side is the elimination of the unfit; as pointed out, even prior to 1933 visits had been arranged to mental homes to bring this out to the peasants. This was made abundantly clear in one *Landesbauernschaft* after the accession. The time of false equality was now over; the RNS would go back to the holy stream of German blood. This would imply rooting out the unhealthy, and the letter spoke of the RM 1,000 million being spent on the inhabitants of asylums or other unfit people who gave nothing in return.[45] Even vocational courses in agricultural schools included visits to sanitoria by way of emphasis.[46] History was invoked when a contemporary writer alleged that ancient Germanic law had included the abolition of the 'less valuable' by castration, sterilisation and the death-penalty.[47]

This type of Social Darwinism was not merely deeply embedded in the NSDAP's conceptual system, it was almost its very foundation, as witness the early law in the Third Reich allowing for compulsory sterilisation. On the land, however, the whole matter assumed a special urgency, as the party was dealing with the very wellspring of the people, by its own definition. The whole *Erbhof* legislation brought the question to a head, as unfitness in the present owner could mean sickly offspring, and the whole point of the law thereby nullified: this possibility did not escape contemporaries. There was plenty of evidence to show that this was indeed happening in some cases: one doctor in the south-west wrote to two *Erbhof* families containing defectives in his own district. He suggested that present policy was actually too liberal, and that no enrolment should be permitted before the hearing of relevant medical evidence.[48]

Hence the constant propaganda barrage kept up about racial hygiene on the land; the picture of a widespread deterioration in the peasantry if the correct policy was not followed was constantly held up before the *Bauern*: in particular it was maintained that the 'less-valuable' were reproducing themselves at twice the rate of the healthy families. Exhibitions were arranged showing the horrific effects, in physical terms, of allowing the hereditarily diseased to reproduce at will, with the comment in one case, 'Wouldn't it have been better if they hadn't been born? ' Apart from trying to prevent the future emergence of peasants physically or mentally handicapped, the RNS exercised some prejudice against the present generation; peasants to whom the sterilisation laws applied were not eligible to be honoured as old-established families, for example.[49]

This was a hard line to follow, but one difficult to avoid in practice once the initial premisses of National Socialism had been accepted. Sterilisation was part and parcel of the basic grammar of the movement's thinking. It would be wrong to suppose that only in the Third Reich was eugenics a topic for discussion, as the whole subject has a long history. Its founder Sir Francis Galton did not, of course, advocate the elimination of the unfit, but in so doing the National Socialists were only really carrying the basic concept of biological improvement to its logical conclusion; in effect they held that furthering the fit is no use without simultaneously suppressing the unfit. Here again cold logic was apparently preferable to humanitarian values, as the latter were believed to be socially dangerous by protecting the weak. The basic Social Darwinism of the party was nowhere more evident than in its agrarian population policy, as evidenced by the use of words like race *(Rasse)*, breeding *(Zucht)* and selection *(Auslese)* which until the mid-nineteenth century at least had never been applied to human beings, only to plants or animals.[50] Here the NSDAP was firmly in the whole völkisch tradition in Germany, and its tendency to assimilate the development of the human race to the laws of natural history. This explains the victory of logic over human values, exactly because in natural history there are not any: Social Darwinism did not possess the word 'ought', except in a purely prudential sense.

The population policy of the NSDAP had a quantitative aspect, as well as being concerned with quality. Since this birthrate issue was so important, indeed fundamental, in the approach to agrarian matters, the belief that Germany's future depended on rural reproduction rates needs to be analysed carefully.

First, since 1921 there had been an excess of births over deaths in Germany, although a decreasing one: this arose through a favourable age grouping in the early twenties which from 1926 onwards had ceased to be the case. In 1934 the country had nearly five and a half million less inhabitants under the age of twenty than in 1910, although the total population was eight million greater. There was a sharp fall in the net reproduction rate, from 1.448 in the 1880s to 0.624 in 1932.[51] Statistics could be produced, however, showing that the birthrate was far higher on the land.[52] Hence the book produced in 1929 by the head of the Reich Statistical Office, which seemed to demonstrate that Germany was dying from urbanisation.[53] In retrospect it appears that a major factor was urban unemployment, which implied that once the work-creation programme was executed the situation would alter. This did not occur to many people at the time, however, who simply assumed that the decline in the cities was due to the alleged defects of their life-style. So Hitler's remarks to the peasants at the Bückeberg in 1933, 'A glance at population statistics shows us that the future of the nation . . . depends exclusively on the conservation of the peasant', were more than mere flattery, as he said much the same thing in private when urging acceptance of the EHG (see page 65). As one historian put it, 'Talk about the peasantry as the life-source of the nation . . . had received a thorough justification from the science of population statistics.'[54] The word 'thorough' is perhaps an over-statement, but certainly the facts pointed to one conclusion at a first reading, which for most National Socialists was sufficient.

What determines family size is clearly a complicated question; economic circumstances, religious belief and no doubt varying personal temperaments all play a part. Here, however, the issue is rather an examination of whether it fluctuated in Hitler's Germany or not, as this may have some bearing on the EHG in itself; had it actually worked against a higher birthrate on the land this would have been truly ironic. The chief researcher into the question in the thirties, Josef Müller, came eventually to the conclusion that family size was dictated largely by personal choice, i.e. the individual peasants' outlooks on life, coloured by a general tendency to be about five years behind the urban areas in birth practices in general.[55] If that conclusion was correct, it would have two highly important implications for the NSDAP; first, its general thesis on the peasantry as the life-blood of the nation was wrong, as the rural birthrate would not remain at its present level for long. True, if five years behind the cities it would always be slightly higher *relatively*, but for the peasantry to keep up the population of the whole country its reproduction rate needed to remain *absolutely* at its present figure. Secondly, if Müller's belief that external factors were of little importance in the matter, the EHG would have virtually no effect; the birthrate would fall in rural areas, but Darré's legislation would not actually make things any worse.

Research by the RNS Staff Office removed even that last crumb of consolation in 1937, when its report demonstrated two things beyond all reasonable doubt; the birthrate on the land was not being maintained as before, and the EHG, especially the prohibition on *Abfindung*, was a contributory factor. In the case of the Kurmark a statistical comparison between selected rural districts and Greater Berlin showed how the post-1933 rate had improved more in the latter. This was surely clear confirmation of the thesis that the return to work of the unemployed would change the urban birthrate. Another LBF made virtually the same point when he included the Depression as a factor in affecting family size.

What really shook the RNS was the apparent uniformity of the trend, hence Williken's own comment that the report was 'shattering' in its account of the 'rapid sinking in the peasants' birthrate'.[56] Another RNS survey somewhat later confirmed for South Hanover-Brunswick what had already been said of the Kurmark. The reproduction rate was lower on medium and large farms than in the case of smaller holdings; as the owners of the former were more likely to be *Erbhofbauern*, it seemed that those farmers most favoured were the least forward in performing their völkisch duty. Apart from this ideological blow there was a technical question. In general, it was the districts with the most advanced farming methods which had the least number of children.[57] This boded ill for the future, because if the only way to keep the rural family large was to maintain its vocational techniques backward then how could the peasantry fulfil its duties in the *Erzeugungsschlacht?* It seems legitimate to suppose that some farmers could afford better methods and equipment precisely because they had limited the size of their families.

Given that smaller families among the farming community were now becoming the rule, then the problem of what had induced this remains; contemporary observers were in no doubt that among several causes, the EHG's provision on *Abfindung* stood out as one of the most obvious. According to the LBF in the Kurmark this was always cited by the *Bauern* themselves when questioned on the matter. Willikens himself accepted this point, as he felt it undeniable that the EHG must lead to a further deterioration. He quoted the need to save in order to be able to pay *Abfindung* from current income, as opposed to the old system of taking up a loan secured against the holding. This would lead to later marriage by the *Bauer* and therefore a smaller family. Evidence of smaller families on *Erbhöfe* than on the lesser holdings seems to offer empirical evidence for this inference, as the smallholdings were often the ones to which *Abfindung* prohibitions did not apply, since they were not *Erbhöfe.* In retrospect it is interesting to note how rapidly a similar point had been made by critics of the EHG, at the time when its general introduction was discussed; a Civil Servant in Württemberg forecast in 1933 that if the *Bauer* had to find *Abfindung* from liquid resources his solution might well be to produce fewer children to demand it.[58]

Of course, *Abfindung* was not the only factor in a complicated situation: by 1939 the RNS itself had gone a long way towards accepting the Müller thesis. The *Bauer* had simply taken a 'false example', as the Staff Office called it, from the towns, namely that family limitation would lead to greater wealth. Indeed the February 1939 report speaks of the peasant now saying 'Another child or a car?' as the vital question to be answered. In this respect it was accepted by the RNS that a desire to help the children rise in the world also had some bearing upon the matter, as clearly it would be easier to assist two in this respect than three or four.

Several other possibly relevant factors can be summarised quite briefly. First, the degree of overwork for the peasant's wife was inhibiting to family size. One contemporary survey showed how income was affected when the wife had to spend that time on caring for children which in a smaller household could be given over to cultivating the holding more intensively. As more intensive cultivation meant greater production, a family of limited size not merely enjoyed larger slices of the cake individually, it could also make a bigger one in the first place.[59]

Two other possible sources of limitation need mentioning, first, the whole matter of low morale among the rural population in the Third Reich. This was cited in Bavaria as a reason for the falling birthrate, as the *Bauern* were so pessimistic about the future.[60] From Hessen-Nassau came the interesting, if not provocative, suggestion that repeated insistence upon the virtues of the Nordic race could in the long run give other Germans an inferiority complex. This in itself might lead them to produce fewer children.[61] Although it may sound a little far-fetched at first, this may have a kernel of truth: there seems every good reason to suppose that the 40 percent or so of the population classified as belonging to other races may have begun to feel in the late thirties that they had less future than the Nordics.[62]

All in all, the question was extremely complicated, as so many different possibilities had to be considered. Whatever the final judgement it can scarcely be denied that *Abfindung* played some part in the decline of the peasants' birthrate after 1933: that such a diminution took place is in even less doubt. Equally the defensive tone of the RNS in public on the matter suggests that the real facts were at least suspected, even before the RNS reports.[63] Incidentally the second of these submitted birthrate statistics to a rather more sophisticated analysis than before. It turned out that the reproduction rate had been falling off in rural areas even prior to 1933, and progressively so. The logical corollary is that the EHG was out of date when it was passed, at least in respect of a population policy. This equally absolves it from causing the decline. Thus the *Abfindung* proviso comes into perspective as a measure tending to reinforce an existing trend rather than initiating a new one. It is instructive to put these facts into the framework of early anti-urbanisation statements, such as the one by Gottfried Feder, 'The big city is the death of the nation', based on the low birthrate in Berlin in 1934.[64]

14

THE MANAGEMENT OF WARTIME AGRICULTURE AND FOOD DISTRIBUTION

Long before the outbreak of war in 1939 the general administrative pattern of agriculture under the NSDAP had been established. Once hostilities commenced an Allied blockade of the Reich was instituted as in 1914, so that inevitably the main accent in the German agrarian scene post-1939 fell on the maximisation of production. Clearly this had to take priority over other considerations, at least provisionally. The main contours of wartime policy can therefore be mapped out fairly easily in one chapter.

In organisational terms the war betokened a shift of emphasis rather than a sudden break. The *Erzeugungsschlacht* and the *Reichsnährstand,* which in other countries might have been occasioned by the outbreak of hostilities, had in the Third Reich long preceded it. Certain other measures taken in secret even before the battle of production itself had begun complemented these. In 1934 a special office had been built into the RNS itself (Stelle für Ernährungssicherung).[1] It had a central staff of ten, and a representative in each *Kreisbauernschaft.* Two million separate farms had already been examined when the war began to find out how war might affect them, whether replacement labour would be necessary, or whether the farmer could be exempted from call-up, etc. In other words, a formidable degree of planning had already been undertaken, so that the changeover to wartime management caused hardly a ripple on the surface of agrarian organisation. This in no way necessarily implies that Germany worked consciously towards a war in 1939, it had merely taken steps to be prepared for one should it arrive. To this end ration cards had been printed as early as 1937.

As the management of food production and distribution had been rather poor in the First World War, and perhaps even contributed to

Germany's defeat, it is hardly surprising that it was felt in the thirties that better preparation should be made for any future conflict. How bad it had been can be seen from a few examples. The Minister of Finance refused to spend RM 5 million on foreign grain in the summer of 1914, because he felt the war would not last long enough to justify the expense. When rationing did become necessary the government issued five million bread cards too many. No central food distribution was set up; rationing was introduced at different times for various commodities, locally managed under the aegis of the Ministry of the Interior, already overburdened. Eventually as the position worsened a central authority (Kriegsernährungsamt) was founded in Berlin, from which the Ministry of Food and Agriculture grew. Price-policy was left to local administration, which resulted in widespread anomalies. Inexperienced officials decided in 1915 to reduce the number of pigs in order to save their fodder potatoes for human consumption. Nearly two-thirds were slaughtered in three months before surplus potatoes which could have fed them were discovered. Not only was a huge potential supply of meat and fats lost, but also fertiliser for the fields. This mismanagement became part of the folklore of NSDAP agrarian leaders.[2]

Thus when hostilities began in 1939 Germany was infinitely better prepared than she had been twenty-five years earlier. There were reasonable grain reserves, amounting to six and a half million tons:[3] the country was almost self-sufficient in output of corn for human consumption. It is true that Germany now had roughly eighty million people to feed as against sixty-seven million in 1914, whilst the agricultural area available was almost identical in size (just under thirty-five million Ha). But higher yields and more mechanisation could provide compensation here. The RNS leaders were entitled to feel that they could cope with the Allied blockade.

The necessary measures to introduce rationing were executed even before hostilities commenced. On 27 August 1939 food offices were set up at two levels, regional and local; the first was headed by the relevant LBF, the second by the KBF, yet another addition to an already formidable list of duties. The same decree placed the RNS formally under the jurisdiction of the Ministry of Food and Agriculture. Ration cards were introduced on the same day for fourteen different kinds of goods, ten of them foodstuffs.[4] Adult consumers were placed in four categories, from 'normal' to 'heaviest worker'. The food offices were themselves divided into two departments, A and B; the first was to oversee farm deliveries, the second to deal with actual distribution. Another ordinance took over all agrarian production by the Ministry, and regulated the amount of food which peasants would be allowed to keep for domestic consumption.[5] Everything surplus would have to be delivered up to the state. In order to co-ordinate imports with internal distribution the Reichsstellen were taken over by the commodity federation already established as part of the RNS. In September the press office of the REM claimed that all necessary preparations had been taken for a long war, and that the management of Germany's food supplies was already superior to that of 1914. An objective observer could hardly have denied the accuracy of this statement up to a point, but certain obstacles still remained. For everything now depended on two things: whether farmers could maintain their production and whether they would keep up their deliveries – obviously two different things in practice.

To take the first problem, Germany was virtually self-sufficient in grain, potatoes and sugar-beet, but fats and fodder were a different matter. Moreover, the relative degree of autarky now attained depended on intensive cultivation, which by 1938-39 was already being threatened by labour losses, as pointed out in Chapter 12. Wartime call-up would obviously endanger the position still further. Fuel and fertiliser might well be in short supply, as other sectors, such as munitions, took precedence for labour or nitrate supplies. Any possible deterioration in the quality of the soil might well be accompanied by the further loss of land to the *Wehrmacht.* Thus at least in respect of available land and labour the outbreak of war merely accentuated existing difficulties, so that here again hostilities implied a change of degree rather than of principle.

The biggest single headache was the prewar problem regarding fodder, and therefore fats supplies. Here again the experiences of the earlier war had engraved itself upon the German consciousness. A lack of animal feeding-stuffs then had resulted in a drastic fall of output in both meat and fats. The 1939 blockade would cut off supplies of oilcake from abroad; the country might risk once again being caught up in the vicious circle where cuts in fodder would entail less livestock, therefore less natural fertiliser and consequently even smaller crops of feeding-stuffs. Now some advantages of the carefully-planned *Erzeugungsschlacht* began to make their presence felt. The drive to be less dependent on outside feeding-stuffs had appreciably increased domestic output in substitutes such as rape and sweet lupins, clover and fodder beets. Just from the harvest of the first-named in 1938 the country had gained 120,000 tons of oil or oilcake, while sweet lupins had yielded 45,000 tons of protein, equivalent to one-seventh of total imports.[6] However valuable these sources were they could never, even with increased beet and potato output, totally relieve dependence on foreign purchases as a glance at the import statistics demonstrates (Table 12).

Table 12. Fodder Imports (000 tons)[7]

	1938	1939	1941	1943
Barley	456.4	397.1	255.7	108.4
Maize, etc.	1895.4	585.8	189.6	149.7
Pod vegetables	66.9	34.1	43.0	23.9
Green and raw fodder	142.5	52.1	219.0	160.6
Oilcake	192.3	62.4	78.8	17.4
Clover	142.6	66.0	19.4	2.3
Misc. products	35.0	25.0	34.5	14.7

Clearly to replace the barley, maize and oilcake imports would need almost superhuman efforts. Moreover, there were several other factors to consider because of the connection between fodder and fats supplies. If grain, beet and potatoes were used directly as human foodstuffs the numbers of pigs would fall, with a consequent effect on fats and meat production. To raise domestic output of these commodities to allow both for fodder needs and human consumption was hard to achieve in wartime because of difficulties over labour. One survey showed that whereas 1 Ha of wheat required 340 man-hours to be brought to fruition, a similar area for potatoes and sugar beet demanded 630 and 740 man-hours, respectively. Once the call-up got under way stepping-up output of the last two would obviously pose problems. True, they yielded from two and a half to four and a half times as much protein per hectare as grain which offset the extra labour.[8] But once manpower got really short even this advantage might partly disappear through sheer lack of labour for root crops.

A further difficulty was that to feed swine on beet and potatoes anyway meant that they achieved an economic weight for slaughtering purposes more slowly, and so the expenses of pig-producers rose correspondingly. The REM therefore raised the farm price for their animals in April 1941, by which time the number of swine had already dropped by between 15 and 20 percent. Their existence was further threatened by a poor harvest in 1941-42 which entailed higher barley prices to make its use as feeding-stuffs uneconomical and thereby reserve it for human consumption. In February 1942 the situation got so desperate that special committees had to be set up for both cattle and pigs, to adapt output to the given fodder situation, led by the appropriate LBF or KBF. Their duties included supervision of the allocation of feeding-stuffs to the peasants, in short, to make the best of a bad job.[9]

The position was improved, however, by a good harvest in the following year, so that producers were not even called on to deliver all their barley to the RNS. Hence there was a temporary slowdown in the fall in pig numbers in 1943, but by 1944 fodder shortages impelled the REM to lower the weight at which pigs should be sold by 10 percent, which temporarily also brought more meat on to the market.[10] But over the whole period of the war the fodder problem proved almost insoluble, chiefly because as the supply of fats diminished, the human consumption of beet, grain and potatoes rose correspondingly, so that by 1944 Germans were eating 30 percent more bread or its equivalent than before the war.[11] This competition for available foodstuffs worked against the maintenance of swine holdings, so that the number of pigs fell by one-third between December 1939 and December 1943. Thus, although the famous *Schweinemord* of 1915 had not been re-enacted, nevertheless over a longer period such a diminution was a serious, although not fatal, blow to German rations. This was particularly the case as pigs yielded such a quantity of fat per protein unit of feed in comparison to other types of supplier, 180 kilos of fat from only 40 kilos of protein fodder.[12]

The country had two other possibilities for fats, dairy farming and oil from plant products. The third source, whale oil for margarine, was lost immediately the war began because of Allied naval superiority; the effect of this can be judged from the fact that in 1938 Germany had been responsible for one-sixth of the world supply of whale oil.[13] That was another heavy blow at supplies.

Dairy farming was by far the biggest single fats producer, but one which again depended ultimately on fodder. On the face of it one answer to this was an extension of the cultivated areas for oil-bearing plants such as rape or sunflowers instead of using animals which demanded feeding-stuffs. But there were two further points to bear in mind: not all soil in Germany was suitable for such crops, and secondly, cows also produced meat and fertiliser as important subsidiary advantages. Ultimately some efforts were made to increase domestic supplies of vegetable oils; in autumn 1939 the price for oilseeds went up by one-quarter, to encourage production. Three years later Backe was still giving their intensified production as one of the most important wartime goals. A further incentive was provided in the form of special deliveries of cooking oil to all peasants who grew the plants which yielded it.[14] But in view of the huge gap in supplies caused by import restrictions vegetable oils were a minor addition relatively to domestic output.

Consequently a heavy burden of responsibility devolved on dairy-farmers to make up the fats deficit. This was virtually another way of saying that it was incumbent to the *Bauer* to do so, as in 1939 peasant holdings up to 50 Ha in size had yielded nearly two-thirds of the total domestic output of milk.[15] And yet it was precisely these holdings which had been hit most by migration and poor prices for dairy-products before the war, so that their livestock was already being reduced in numbers. Between 1933 and 1939 nearly one-quarter of all male helpers had left the family holding (i.e. as distinct from hired labour). The financial disequilibrium between the price-indices of animal products and vegetable/crop products in 1938-39 amounted to 15 percent. In other words, the price-structure was least beneficial to those people whose services Germany now needed most.[16] Backe tried to excuse this in 1942 by emphasising that fodder crops had had to be given good prices to encourage home producers and so lessen dependence on foreign sources, which he acknowledged had hit hard at the peasants (as domestic feeding-stuffs were dearer).[17] But an obvious remedy would have been simply better prices for livestock and dairy produce, as the RNS leaders had demanded at the time. Hitler's obduracy on this point was going to cost the country dearly in terms of fats supplies, exactly as Darré had foreseen. In his memorandum of January 1939 (see page 195) he stated quite explicitly that livestock herds were already being reduced and that things would be even worse in wartime for the peasants. Thus although the *Erzeugungsschlacht* had prepared Germany to some extent for wartime, in one respect at least a little more concern by the Führer might have obviated considerable difficulties later. In other words, the tempo of rearmament, and the price stability entailed by it, had worked against more thorough preparation for a long war. This may well be interpreted as meaning that Hitler only intended to fight a short one if at all.

Whatever the final truth of the matter, the country was going to have to increase milk-production to cover the fats gap and that at a time of fodder and labour shortage. Consequently a special *Erzeugungsschlacht* was launched in 1940 for milk alone. A committee was installed in Berlin with branches at *Land* and *Kreisbauernschaft* level and in individual dairy firms, where the OBF was a member. Competitions were introduced with certificates for the best performers and each committee was responsible for ensuring the best possible use of fodder. The winners got their awards from Darré in person in 1941, when he took the occasion to spur them on to even greater efforts by painting a gloomy picture of what was in store if Germany lost the war.[18]

Eventually financial incentives had to be employed. A struggle took place again in the corridors of power over milk prices before they were at last raised.[19] A premium was introduced of 4 Pfennigs per litre (or 1.2 Pfennigs per fat unit) on all deliveries above 60 percent of that of the average holding. By 1942 some RM 1,000 million had already been paid out to producers in this way.[20] The result was that five years too late German peasants began to fare very well indeed financially. In 1938-39 the price-index for animal products stood at 111 (1909-14 average = 100), but four years later it had climbed to 128, or almost exactly 15 percent. This was ironic in one sense, as the NSDAP had always pilloried the 'System Time' for its policy of state subsidies to agriculture rather than fixed prices. Now under the pressure of circumstances it returned in effect to the same concept.[21]

In defiance of all financial inducements and cajoling the position over fats remained difficult, despite an actual increase in the size of the national dairy herd. The cutting-off of other sources of supply simply meant that diary farmers had to cover an even bigger 'Fats Gap' than prewar. In fact, by 1943 the country was producing 60 percent of its own butter, as compared to only one-third in 1939. But wartime rationing had reduced fats consumption per head, which in 1939-40 was 70 grammes weekly less than in the previous year, although the fall was chiefly in margarine which underlines wartime production achievements. On the other hand many individual producers were still lagging woefully behind the national average, hence the desperate appeals by one Gau with the slogan, 'The last milk reserves are decisive' with reference to the effect of fats shortages in 1918.[22]

A subsidiary result was a greater control over butter-making itself. From the beginning the RNS had tried to make the peasant give up making his own butter and instead deliver milk to a dairy where production methods were more efficient, so that far more could be turned out from the same quantity of milk. So whereas nearly half of domestic supplies had been made on farms in 1932, the proportion fell to under one-third in 1935, and only one-sixth by the outbreak of war. Before the conflict ended virtually all butter was being produced in dairies or factories.[23] This may well have been partly due to shortages of labour on the holdings, but the outcome was the same in that the country as a whole gained by this transfer, which justifies the RNS pressure on the peasant to send milk to the dairies.

As in the case of milk committees were organised to boost domestic meat output, again via higher prices. The index for cattle for slaughter went up thirteen points between 1938 and 1943, the chief jump being in the last year, 1943-44. Stocks of cattle seem to have been fairly constant throughout the war, at between nineteen and twenty million head. Unlike pigs they did not compete with human beings for the same food, so that the maintenance of the herds was due primarily to increased fodder harvests. Not only did the amount of land used for hay stay the same, the cultivation areas of other types of domestic feeding-stuffs, grain, lucerne, sweet lupin, etc. actually grew 50 percent in wartime, having already doubled between 1933 and the outbreak of war.[24] This success story was partly due to foreign labour, of which more later.

Once war had commenced all available food sources had to be utilised to the full, so that Germany's one and a half million allotment holders were drawn into the battle. The REM specifically recommended rabbits for them as a way of augmenting meat rations, as they could be fed on kitchen refuse. If an extra piece of land was available then a pig could be kept. Theoretically this all sounded promising, but snags occurred when householders began to give their rabbits winter greens. As again these were required for human nourishment it was decided to give out certain specified vegetables only on ration cards, although the measure was not to be published in the daily press. On balance, however, allotments undoubtedly helped to ease the rigours of rationing, especially as pigs were kept so often, although meat produced in this way was for the householder only. Then in 1943 the REM announced that henceforth permission would only be given in cases where persons had sold, or would sell, some of the meat rather than consuming it all themselves. Three months later regulations were tightened still further.[25]

On the whole in wartime the NSDAP's agrarian policy was virtually a continuation of the *Erzeugungsschlacht.* There was roughly the mixture as before, appeals to duty such as Goering's in February 1940 which paralleled his earlier one of March 1937. Equally, there was a gradual tightening of controls over deliveries which had really begun with the Four Year Plan. Backe's arrival in full power in May 1942 intensified this trend due to his tendency towards state intervention *(Etatismus)* as compared to the perhaps more romantic Darré and 'peasant policy'. Doubtless, apart from his knowledge of the Soviet Union, this made him a more useful man to Hitler than his predecessor had been. What the Führer required in wartime was efficient pragmatism rather than dreams of 'Blood and Soil'. Of course, ideologically Backe was a National Socialist too, but in personal terms he was probably more efficient than Darré, and more hard-headed. It has even been suggested that the latter occupied himself only slightly with wartime production.[26]

So agriculture underwent increasing supervision from the beginning of the RNS in 1933. The milestones on this road included the *Erzeugungsschlacht*, the Four Year Plan, the takeover of the RNS by the state in 1939, and finally Backe's arrival. This must not be pressed too far, as the wartime scene contained the carrot for the farmers as well as the stick, as had the peacetime one. Cajolery, flattery and appeals were accompanied by competition, premiums and higher prices. In this respect Backe simply followed Darré's earlier policy. (Labour regulations will be discussed later.) This was not necessarily any aversion to the use of further authoritarian methods as rather to the feeling that they failed when employed.

A good example was food deliveries. The 1939 regulations in effect compelled producers to hand over everything surplus to home consumption to the state. And yet there is much evidence of black marketeering after this decree. Indeed penalties were being announced for such activities by the following year. By 1942 farmers were being ordered to hand over their entire bread-grain harvest under pain of penal servitude (*Zuchthaus*) for non-compliance, plus fines without legal limitations. Similarly the state was empowered to take animals where necessary, in cases even where they were not directly required for slaughter.[27] This was a formidable list of powers and yet, even as late as January 1943, Backe had to say specifically that to ensure proper rations for everyone food had to be delivered as well as produced. Evidently coercion had not produced the desired results, even though there was a catalogue in every *Kreisbauernschaft* with a card for all individual holdings over 5 Ha. The officials knew pretty well how much each peasant should produce, and yet in 1944 they were still appealing to them to do their duty. Figures were printed to show how much more could be delivered if peasants refrained from direct dealing with the public. '*Bauer* become hard' pleaded the party in one Gau.[28] It is inconceivable that such language should have been used if threats alone had worked.

There is no doubt that a Black Market flourished in wartime Germany. For instance, in August 1942 a sudden growth of holiday-makers in rural areas gave rise to the suspicion that they were seeking food illegally. Often it was exchanged against other goods and services, rather than bought outright. In January 1944 a regular tariff existed in two South German Gaue, where a packet of pipe tobacco equalled 500 grammes of butter, or one egg for a cigarette. One East Prussian peasant acquired a fantastic amount of new clothes in a year despite rationing, the implication being that barter had been responsible.[29] Sometimes traders rather than peasants were the culprits, so that the cattle and milk federations were given powers to shut down any business in their sectors which had infringed delivery regulations. Later the same year, poultry farmers who failed to achieve the modest yearly delivery quota of sixty eggs per hen were threatened with the same type of measure. As the announcement frankly admitted normal punishments were no longer effective, so that the offender could now be forbidden to keep poultry as well as fined.[30] All in all, despite Hitler's threat at the beginning of hostilities that anyone who tried to turn the situation to his own personal profit would earn only death, it appears that regulations failed again to obtain full control over food distribution.

Apart from shortages of feeding-stuffs the main factors inhibiting even greater maximisation of output in wartime Germany centred around mechanisation, fertilisers and the whole question of the amount of labour available. The relative success in the *Erzeugungsschlacht* prior to 1939 had been achieved by intensive cultivation so that clearly anything militating against continuation would have serious repercussions.

The position for fertiliser was reasonably satisfactory. There were supplies of all types of artificial manure available in Germany except phosphorus. To ascertain soil conditions exactly a survey was undertaken by the REM in March 1940. It was obviously vital that the fertility of the land should be maintained, especially as natural manure might be lacking to some extent. In 1940-41 overall deliveries of the artificial kind were much the same as two years previously: shortfall in nitrogen was compensated by extra potash output. At the beginning of hostilities the REM had over-reacted slightly to a possible future nitrogen shortage, as priority would be given to its use in the manufacture of explosives, and a cut of one-quarter was made in the supply to create a reserve, and almost at once restored. As the conflict progressed shortages developed inevitably, partly through the blockade again; in 1939 Germany bought a million tons of raw phosphate abroad but four years later it could obtain only about one-sixth of that figure.[31] By March 1944 it was being admitted that a general lack of fertiliser was making it hard to keep up the quality of the soil.[32] Other types of manufacture had priority, which accentuated the problem of the blockade, and the Allied air attacks on the communications system added to the problem.

Accompanying this deterioration was an abrupt halt in the mechanisation of German agriculture, obviously at a time when it could have done with an acceleration. Ultimately this had two principal causes; iron and steel were needed for munitions, but quite apart from that there was the question of fuel. Since the war intensified demand there was little left for civilian requirements, even in the agrarian sector. In summer 1940 it was ruled that a new tractor would be made available only in those cases where the old one was irreparable. The need to save petrol first limited use of such equipment to ten days monthly, two years later, and it was decided to introduce a substitute fuel from wood *(Holzgas)*.[33] Simultaneously the production of tractors operating on liquid fuel was cut by a half, and it was announced that from 1 July 1942 only vehicles using *Holzgas* would be delivered to agricultural purchasers. The regulation pointed out that as the new type of machine required an ample supply of dry timber, holdings with plenty available would be preferred. This would clearly discriminate in favour of the larger farms to the detriment of the peasant holdings, although an estate was more likely to have possessed a tractor anyway. In the final analysis this hardly mattered, in that production of farm machinery fell so sharply that in the end almost no one was being supplied.[34] In a sense this was rational, despite labour shortages; there was little purpose in turning out machines for which no fuel existed if the iron and steel could find a use elsewhere. Incidentally, *Holzgas* was far from efficient as a substitute which clearly affected agricultural production. Moreover, both the export of German machinery and of fertiliser during the war had an effect on domestic food output; RM 172 million worth of equipment went east in three years including 7,000 tractors and 250,000 steel ploughs.[35]

The cumulative result of all this clearly threatened yields per hectare, especially as prewar rural migration had been largely offset by greater fertiliser use and increased mechanisation except perhaps in 1938-39. In wartime such compensation was no longer available.

It was therefore more important than ever to maintain the labour force. This received public recognition when the *Wehrmacht* discharged, at least temporarily, many peasants and specialised landworkers as soon as the Polish campaign was over. As the conflict gradually lengthened such concessions were no longer available. In 1942 the *Wehrmacht* had nearly eight and a half million men, the following year well over nine million. The agrarian sector could not alone be exempted from this mass conscription, however necessary its products, as statistics show. On 31 May 1939 there were almost five million adult males employed in agriculture and forestry, but on the same date four years later approximately three and a quarter million. It is true that many peasants or their sons were spared military service, just over 600,000 at 31 May 1943. This was welcome when it was granted but apparently caused some friction in the villages, and rumours were not lacking that the ones spared owed this to their relationship with the local OBF.[36] Doubtless this hardly contributed to morale on the land, but this reserved status *(UK Stellung)* for some was probably the outcome of the prewar preparations already referred to (see page 221).

As before, when shortages were threatened the first step was recourse to legal sanctions. In the first week of the hostilities a decree tightened up labour control in general.[37] One interesting clause decreed that no one could be dismissed without official permission unless his last employment had been in agriculture. This may have been intended as an incentive to employers to let former landworkers go (who had illegally left agriculture) as it did not involve contact with the labour exchange if they wished to do so. Probably conscious that this proviso would make little real difference in the long run, Goering appealed in February 1940 for neighbours to lend a hand on the farm where it was needed.

As the call-up bit deeper into the ranks of peasants and their sons it became clear that even volunteers would not suffice, and direction now began to emerge, step by step. The first was in April 1941 when the OBF was empowered by decree to direct (agricultural) labour within his domain where he found it necessary. He could now put a peasant's son to work for a specific task on another person's holding, but could not direct non-agricultural labour. By the following year the position necessitated a further tightening of the screw; in March a special order from the Führer established a Labour Commissioner to co-ordinate supply. In the same month full-scale direction was introduced for the agrarian work-force.[38] From then on all persons either on the land or in rural areas could be bound to the OBF for a certain period of time to perform stated tasks for regular wages. He was to draw up the list of those to be conscripted in conjunction with the branch leader of the party and the Bürgermeister, including especially any persons formerly employed in agriculture. This measure was taken on the authority of the February 1939 regulations (see page 201). In other words, provision for labour conscription had long preceded hostilities. Anyone refusing the allotted tasks could be reported by the OBF to the labour exchange and faced a possible penalty of fine or imprisonment or both. If they accepted, however, they got extra rations. This inducement suggested that the NSDAP did not wish to apply the stick unless they had to, to avoid friction and thereby lower morale. Obviously sugaring the pill by giving extra food would help to ensure that an unwilling work force might be dissuaded from simply slacking through resentment.

In some cases the conscripted persons, if already in agriculture but being redeployed by the OBF, were already classified as self-suppliers for rationing purposes. Where such a person refused direction he was to be reported to the Food Office, which would reduce him to the level of a normal consumer. This meant that a peasant so re-designated would have to deliver all the food he produced instead of being allowed to keep a certain amount for domestic supply. Additionally, he would no longer have the right to keep a pig for his own consumption. Gradually the net was being tightened even over the peasant proprietor himself. The final twist was applied in February 1943; after an LBF's meeting at the REM it was decided that henceforth all owners or lessees who had been militarily exempted were now to stand automatically at the disposal of the appropriate KBF.[39] The *Bauer,* having first been reduced to a tenant in effect by the EHG was now no longer even permitted to decide to remain solely on his own farm if the KBF decided he would be more useful elsewhere. It has to be admitted, however, that this was a wartime measure and that labour conscription was applicable in all countries engaged in the conflict. There is no reason to suppose that this and similar regulations would have remained in force had Germany won the war.

The peasants placed at the KBF's disposal were allocated usually as advisers on farms where the owner had been called up leaving their wives to cope perhaps without assistance. They were now provided with a kind of godfather *(Hofpate)* in the form of another experienced manager. Such schemes began in April 1943, and were claimed at the time to be proving of great value for the farmer's wife.[40] There does not seem any reason to doubt this, as at a time of labour shortage and increasing difficulties over fertiliser and machinery the arrival of an experienced supervisor may well have seemed a godsend to an overworked woman with her own family to look after.

So by mid-1943 direction of domestic labour was in full swing. Alone, however, this could scarcely suffice to replace the nearly two million men conscripted from the agrarian sector. As before the war other sources were necessary. As early as 1941 a directive from Hess made all schoolchildren upwards of ten years of age available *(einsatzfähig).* Until they were fourteen they were allocated light work only, but then became treated as normal workers *(volleinsatzfähig).* In the case of urban children the age limit was sixteen. Weekends and school holidays were especially to be utilised. By 1942 juvenile help was extended by the Ministry of Education, which closed all agricultural training schools on 15 May in order to release pupils for summer work on the land: teachers in them were also expected to participate.[41]

All the prewar volunteer schemes for youth continued unabated. The Labour Service's female section was a potential aid to the overworked peasant's wife and on 4 September 1939 plans were mooted to raise its strength to 100,000 girls by calling-up unmarried ones aged between seventeen and twenty-five where they were urgently needed in agriculture. Numbers were augmented still further by a special Führer decree in July 1941, and the service period extended to one year. The strength was already 130,000 and was to be raised still further by another 20,000. Ten months later there was such a lack of labour for the root crops harvest, so desperately needed for fodder, that the terms of employment were again amended. Henceforth a girl called up in spring would do seven months active service on the land and five as auxiliary war helper *(Kriegshilfedienst)*.[42] For those conscripted in autumn the conditions were reversed. This ensured experienced labour for the harvest by obviating the use of newly-arrived girls in September. All would be gathered in before they arrived.

Even if the individual girls were relatively unskilled they could still carry out domestic chores such as housework which relieved the hard-pressed wife, or lend a hand with the kindergarten duties performed by the NSV on the land (see page 189). In dairy farming areas they were expected to learn milking as well, either on a special course or through itinerant instructors who went round the various camps in which the girls lived. In one mainly rural area in North Germany there were as many as seventy-seven such camps in April 1943.[43] The entire scheme was well-organised and useful as a source of auxiliary labour.

Important as all these replacements were they could never in themselves have made up for conscription of male labour. Fortunately military success provided a remedy in the shape of prisoners of war and civil labourers from the east: without them agricultural production would virtually have collapsed. At 31 May 1943, 1,750,000 German males had been pulled into the armed forces from agriculture, but against this 700,000 prisoners and 1,500,000 foreign workers were already at work on the land. Consequently the labour force remained quantitatively speaking relatively stable. If the 1939 frontiers are taken as the basis, there were in that year just under eleven million persons in agriculture (excluding forestry), a figure which by the following year diminished by about half a million. Then in 1942 there was a return to the 1939 total, which remained steady right through to almost the end of the war.[44] The sudden increase in 1942 was the effect of the Russian campaign, which not only brought more prisoners but enabled more civil labour *(Ostarbeiter)* to be recruited.

At first sight then agriculture was well-supplied with labour after 1939; indeed it now seemed to be better off in that respect than prewar, ironically enough. But it had to be remembered that the eastern workers were not well-treated in general: their remuneration was well below that of a German according to a special scale which existed for them. Their wages were calculated as a certain proportion of that which a native worker would get for the same job. Equally their treatment often left something to be desired, to put it no stronger. Here there should be no hasty generalisations, as obviously this depended largely on the individual peasant with whom they were quartered. The authorities threatened heavy punishments for Germans who treated the *Ostarbeiter* too well but it was never easy to lay down exact instructions on the matter. In one Gau it was pointed out that they were neither fellow-countrymen nor political deportees; presumably the party wanted a reasonable compromise between these two implied attitudes. But clearly there was no conceivable question of equality. As the war progressed and more brutal methods of recruitment were utilised to acquire eastern workers their goodwill diminished and acts of sabotage took place.[45] Qualitatively *Ostarbeiter* could not replace willing workers. But not all foreign workers were from the east; agreements with Italy enabled the Reich to acquire by 1941 some 50,000 workers for the agrarian sector from that country.[46] Moreover, even if qualitatively eastern labour was sometimes lacking in enthusiasm, the lower wages often permitted a poorer peasant to use a Pole or Ukrainian where a German might have been too expensive. Some holdings may therefore actually have been better off in this respect than formerly.

Prisoners of war formed another cheap labour supply. As they were often used on land-improvement it tended to be the state which benefited here, as captives received only 60 percent of a German worker's wages for the equivalent tasks. When they were allocated to an individual holding the proprietor had to pay the *Wehrmacht* for their upkeep (RM 1 daily), or provide it free.[47] This again does not sound a bad bargain for the farmer. Clearly the 700,000 POWs in employment on the land in May 1943 were a windfall in some respects, but again their enthusiasm was understandably somewhat limited as they were being used to help what was to them an enemy power. Consequently it is hardly a surprise to encounter so many complaints in reports about their low output. Another disadvantage was the effect which they exercised on German morale, which appears to have been particularly the case with British POWs. The certainty which they displayed about an unfavourable outcome to the war for Germany was not conducive to raising enthusiasm for the conflict in the local population.[48]

Although numerically important, the POWs were a mixed blessing for the agrarian economy, as also were the eastern workers. From a nationalist standpoint the sight of over two million foreigners in German villages, the majority of them Slavs, was hardly welcome. Even prior to hostilities the party was worried about the phenomenon of imported workers replacing Germans on the land (see page 198). A few years later things had reached such a pitch that in some places German women were afraid to go out alone after dark because so many foreign labourers were in the vicinity. What made the whole thing worse was that no one could pretend any longer that Germany would be able to dispense with their services even if she won the war. This was a political headache even by 1941, when one Gau had outside help from thirty different nations in all, amounting to 100,000 individuals. The local NSDAP admitted frankly that there was now little hope of reversing migration by bringing Germans back from the cities, hence the acceptance of foreigners as a feature of the rural landscape for the foreseeable future.[49]

This apparent fatalism was induced by years of failure to check land flight, which continued even in wartime. True, there had been a slight slowing-down in the pace of migration but even in 1941 60 percent of all rural male school leavers in one local survey had gone from their villages. Possibly the attractions of less Allied bombing and more food played some part here which would obviously no longer apply when the conflict finished. In 1940 it was calculated that agriculture required 310,000 apprentices in all its diverse trades combined to maintain its labour force. Yet four years later only 60 percent of current needs were being achieved.[50] The wider implications of such facts will be examined more fully in the Conclusion; the main point here is that the widespread use of foreign labour on the land, in combination with the migration of young Germans to the cities hardly promised a bright future for the concept of the rural population as the life-source of the German people.

Furthermore, in qualitative terms in wartime relatively skilled German labour was often either called-up or sent to occupied countries. As the war developed many administrators had to be sent east or to territories taken over elsewhere. The further the *Wehrmacht* advanced the more trained personnel were removed from the domestic agrarian scene. Equally the 427,000 Volksdeutsche resettled in Poland by 1 April 1941 required technical and managerial assistance. Most of them were on the land and needed help in adapting themselves as not all were necessarily efficient farmers. This was an additional call on technicians and administrators from the homeland, who had to be sent to newly-annexed areas in Poland as an advisory service.[51] In the Soviet Union German agricultural administrators became an especial target for partisans, and from 1942 onwards almost every issue of the official government gazette for the land *(RMB)* carried the names of those who had fallen on the eastern front. This was a heavy drain on internal resources, in a country already needing all the technicians and labour it could get for its own agrarian management. In 1937-38 lack of technical assistance had already cut back some land improvement in Germany itself.[52]

The overall picture of the management of wartime agriculture is very much one therefore of a constant battle against obstacles of one kind or another. Most prominent among them was that of a diminishing work-force, at least in qualitative terms plus the loss of overseas fodder through the blockade, difficulties over fuel and machinery, and finally of fertiliser. All these combined to reduce yields per hectare in crop products, and diminished the stock of swine by one-third in four years. Obviously while they were being killed off the meat ration could be kept going, but their loss eventually meant less fats, which as whale-oil for margarine was unavailable was a grave matter.

The most disturbing aspect of the difficulties was in respect of crops, including fodder, as the soil gradually declined in fertility. This emerges very clearly from an examination of yeilds per hectare, as shown in Table 13.

Table 13. Yields per hectare (in double zentner) in selected groups[53]

	Rye	Wheat	Summer barley	Winter barley	Oats	Potatoes	Sugar-beet
1933	19.3	23.7	21.0	26.0	21.8	152.6	282.1
1939	19.8	23.5	21.3	25.0	21.6	182.2	333.4
1944	16.8	21.4	17.9	24.0	18.2	142.9	251.7

As fodder yields fell in a roughly similar ratio to that of sugar-beet, the decline was catastrophic. However, the biggest reduction only came right at the end, in 1944 itself. Until then harvests had been more or less constant, depending mainly on the weather. If the grain deliveries for individual *Landesbauernschaften* are taken as the criterion there was a big drop at the beginning of hostilities, a relative recovery by 1941-42, a very good year in 1942-43 (at least for some types of corn) and again the following year, and a final decline in 1944. This was partly due to cumulative deterioration of the soil but by then Allied bombings were also disrupting communications and lowering rural morale.[54] Therefore it becomes hard to assess exactly what part soil deterioration played in producing lower yields.

In November 1942 Backe set the agrarian community five goals to attain; the maintenance of grain supplies, the maximum possible output in potatoes, green vegetables and sugar-beet, new records for oil-bearing plants, more milk, and more pigs. The last was never achieved but until 1944 farmers had by no means done badly in respect of the others, despite the handicaps of fertiliser and machinery shortages. Foreign labour compensated for a further thinning in the ranks of the domestic workforce after hostilities, quantitatively if in no other way. But ultimately nothing could keep up yields, especially once the Allied air offensive developed. Accompanying the drop in output per hectare was a decline in sown areas in grain as well, which heightened food difficulties by 1944: in this respect it has to be remembered that although the labour force remained constant over the period 1939-44 it had been inadequate in the first year anyway. Eventually farmers had no option but to go over to a less intensive cultivation, despite all the labour direction and makeshifts such as *Hofpaten.* This is clear from the statistics given in Table 14. Fodder, beet and potatoes were kept up very well, but grain was well down, so that by 1944 the position was becoming serious, particularly as outside sources of foodstuffs were being given up because of military withdrawal.

Table 14. Sown areas (000 Ha) in various crops[55]

	Rye	Wheat	Winter barley	Summer barley	Oats	Potatoes	Sugar-beet	Various fodder
1933	4524	2431	271	1315	2905	2718	304	513.0
1939	4223	2105	425	1243	2820	2834	503	1001.7
1944	3851	1781	301	878	2438	2704	543	1498.6

Quite obviously, although the occupation policies of National Socialist Germany lie outside the scope of this work, no description of German agriculture in wartime could possibly be complete without reference to the latter point. For the four years from 1940 onwards the Continent lay almost wholly at German disposal, for food as well as for other products. In the west there was a well-established dairy farming industry in the Netherlands and Scandinavia as well as the rich farmlands of France to draw on. In 1939 Holland, Norway and Denmark all had far more cows per 100 Ha of agricultural land than Germany.[56] As to the grain areas of the east, they seemed to offer mouth-watering possibilities. One RNS leader assured the public in 1940 that if only they were as intensively cultivated in the Soviet Union as in the Reich, output could be multiplied by a factor of five and a half.[57] The policy directives of the agrarian group on Economic Staff East were clear that the major objective for them was to remove the grain surplus in the Ukraine, hopefully estimated by Backe at some seven million tons annually. The Soviet Union was to provide so much meat as well that the Führer wished to use it to restore previous ration cuts made in the Reich and indeed by autumn 1941. The one great advantage which the east had over the west in National Socialist eyes was that whereas both areas had foodstuffs which would enable the Reich to withstand a blockade, the inhabitants of the latter area had themselves to be decently fed, whilst racial ideology ensured that Slavs did not inconvenience their occupiers in a similar fashion. In addition the Reich needed western industry for the prosecution of the war, whereas manufacturing capacity in Russia was held to be less essential. In future, therefore, grain surpluses from the Ukraine would come west rather than being as previously, fed to the Great or White Russians. As its industry would be extinguished in this process, it would no longer exist to threaten west European manufacturers, who would then have a new market in the Ukraine, unhampered by competition from other parts of the Soviet Union.[58]

Unhappily expectations of booty turned out to be higher than actual receipts, partly due to the exigencies of war (e.g. the activities of Russian partisans) and partly to sheer mismanagement. Indeed, as the Soviet Union already had a commercial treaty with Germany the latter, in 1941 at least, would have received more Russian grain without an invasion.[59] This, however, is not to say that there was no ultimate food profit obtained. Considerable supplies of meat, grain and fats were taken from the occupied countries but despite the impressiveness of the figures which follow, several important qualifications need to be made (Table 15). First, under imports is shown *Wehrmacht* consumption in the occupied countries, so that the civil population in Germany gained little. Moreover, by attacking Russia the Reich lost ordinary treaty supplies from her and in wartime purchased less food in south-east Europe, so that two peacetime suppliers were replaced by one enlarged wartime one. As can be seen, Germany obtained substantial supplies from the east in general, which were an important additional source of foodstuffs at a time when they were most needed. But these figures include army consumption in the Soviet Union, which was inevitably on a very large scale. In other words, the Ukraine's chief advantage was circular: it helped to feed the troops sent to occupy it.

Table 15. German Net Imports in selected years (000 tons) from East and South East Europe, including Wehrmacht consumption[60]

	Meat	Grain	Fats	Potatoes	Sugar
1937	142	1072	94	11	–
1938-39	160	2458	61	9	–
1942-43	436	4567	260	2895	327

In so far as meat and fats were concerned, the ultimate picture varies only slightly. Before the war the country supplied 92 percent of its own meat, and even wartime imports came mostly from western sources; for example, 422,000 tons out of 741,000 tons in the fourth year came from France, Denmark and the Netherlands. Furthermore the foreign workers and POWs in the home country (eight million by 1941) required 700,000 tons of bread grain, 90,000 tons of fats and 150,000 tons of meat annually. The net gain for the civil population was small in comparison to total domestic production. Similarly, although some fats came from the Soviet Union later, their imports as a percentage of total output is higher in 1939-40 than in any subsequent year. Whereas in 1938 and 1939 the country obtained 92,000 tons of butter abroad, imports fell to two-thirds of that total in 1941 and to less than half of it two years after that.[61]

Additionally, there were still German food exports during the hostilities. This originated partly in a policy of exchanging required industrial goods from the occupied west against foodstuffs. In 1933 food constituted 4.6 percent of Germany's exports, a proportion which dropped to 1.2 percent six years later. It then rose steadily every year up to 1943 when it actually attained 6.8 percent, or half again that of ten years previously. Another factor here was simply the pattern of German military conquest, since not all the territories acquired were agriculturally valuable. A classic example was Bohemia-Moravia. The Reich had to agree to cover its food deficit, as it needed its industrial goods. Consequently between 1 August 1940 and 31 July 1941 it sent 20,000 tons of meat and 30,000 tons of fats to the Protectorate.[62] This obviously has implications for any study of Hitler's foreign policy to which reference will be made later.

In sum, whatever rosy expectations may ever have been entertained about the Ukraine in particular, in reality Germany benefited little from the occupation, at least in terms of foodstuffs. Here the effect of the Allied blockade, which by extension became one of the whole Continent after Britain had been driven out in 1940, was important. Most west European countries had become dependent upon overseas fodder for their domestic agriculture;[63] deprived of that, their output had to suffer as much as did Germany's for the same reason. Hence the need of several occupied western countries for assistance from Germany, so that the Reich often used foodstuffs from other areas to obtain industrial goods from the west.

So whereas in 1941 a normal consumer in Germany received 2,400 calories daily, in 1943 he got only 2,200 and in early 1945 only 2,000.[64] But this conceals increased consumption of grain and potatoes and less of meat and fats. The original meat ration of 2,000 grammes monthly was maintained until 1941, when it fell to 1,600 grammes until 1942, and was mostly even smaller for the last two years of the war. Fats were also cut in April 1942 from 1,000 grammes and never subsequently rose above 900 per month. Against this, milk, bread and potato rations were held more or less constant until the last twelve months. The general picture, in other words, is of a generally reasonably fed community whose diet progressively deteriorated in quality, rather than one living off the fat of the occupied lands.[65]

Finally, there is the matter of settlement outside the Reich in the war. The execution of this task was placed in the hands of Heinrich Himmler, who originally had to deal with the re-settlement of ethnic Germans who could now be brought back from other countries in to the newly-annexed slice of Poland. His new office (RKFDV) was set up in October 1939 although the decree was not actually published until 1944. This was rather a slap in the face for Darré, whose protests over the preferment of his erstwhile friend were in vain. About all that the RBF could secure was that all holdings taken over from Poles and Jews were put under the administration of the REM, even those confiscated by Himmler.[66] Probably this downgrading was a further reflection of Hitler's diminishing confidence in Darré, once so high in his favour.

Himmler was convinced that the area could only flourish if the settlement was planned and above all balanced. The east must call on artisans and traders and professional people as well as on the peasant, who would be the foundation-stone rather than the complete edifice.

By November 1940 the SS chief was in a position to issue general guide-lines, setting out the pattern of future Germanisation in some detail.[67] All farm sizes were included, with a peasant holding of 25-40 Ha preferred, as soil conditions precluded anything smaller, except for specialised products such as fruit. A few estates would be accepted employing German farmworkers, who could by diligence rise to become *Bauern* themselves. Ideally each village would have thirty to forty small plots, fifty to sixty normal *Erbhöfe* and seven to nine larger ones, with one or two estates in a ratio of 10:55:23:12 in percentage use of available land respectively. Characteristically Himmler referred to the peasant holdings as 'hides' or 'big hides' *(Hufen-Grosshufen),* a deliberate revival of the terminology of the thirteenth-century Germanic colonisation of the east. Such villages would be grouped around a larger one *(Hauptdorf)* which would serve as a kind of local centre.

Similar schemes were mooted simultaneously in other quarters; one contemporary work went into details over village sizes, a school system, even the proper balance to be struck between industry, commerce and agriculture. Actual models of communities, ideal farmhouses, etc. were produced in the course of the planning.[68]

Land had been acquired and plans drawn up, but where were the settlers? This became especially acute as later conquests widened the living-space available, as everyone concerned with the matter was absolutely convinced that only occupation by Germans could really transform the conquered land. However vague the Führer might be about the actual numbers to be moved east, his subordinates were with him in principle on this. As one commentator phrased it, 'Only the binding together of sword and plough can make a land really German.'

Neither Civil Servants, a remote ruling class, traders nor the church could do this, only the German peasant tilling the soil, which implies a mass movement. Hence the appeal in an SS publication that the east was calling the German people. The concept was above all related to the future, since a growing rural population could be rooted there over centuries of demographic expansion.[69] Time would in any case be needed to eradicate the last remaining tinges of individualistic liberalism among the German people.

Because of this the chief hope of future settlement had to lie in German youth, a conclusion which the SS faced squarely in discussing availability of the necessary human material. It quoted its own leader to the effect that a new spirit had to be created in German youth. So once Poland had been defeated the propaganda floodgates opened with a vengeance; the message was always the same, and could be summarised quite simply as 'Go east, young man! ' Axmann of the Youth Service was first in with the call for volunteers for the Land Service in the east.[70] Once units of the Hitler Youth had been formed in the newly-won slice of Poland, the idea of making the east an essential part of the national future was hammered home at every propaganda evening.[71] The area became in the midst of war the talking-point of almost every official publication. In January 1941, for example, the *National-sozialistisches Monatsheft* devoted almost its whole issue to it. A later number even dragged German history in to buttress its arguments by evaluating the eastern policies of the medieval Hohenstaufens.[72] By 1943 it was claimed that 18,000 young people had gone to the territories.[73]

The indoctrination was planned to commence even in the primary schools themselves. Along with the Hitler Youth they were seen as the seed beds *(Pflanzstätten)* where the conviction must be sown that Germany's future lay in the east. Use was made of the famous passage in *Mein Kampf* where Hitler had written of the need to arrest the eternal German pull towards the south and institute a new *Ostpolitik.* By April 1942 the party's cultural office's collection of popular, approved songs to be particularly practised in the coming months in the primary schools included two quite obviously geared to this, entitled respectively 'Dost thou see the dawn in the east? ' and 'Forward to the east! ', subtitled 'The Russia song'.[74] None of this necessarily proves that Hitler had always intended to attack the Soviet Union or Poland, but in conjunction with so many of his statements it does rather suggest that under his direction Germany was literally re-orientating itself on a long-term basis.

Equally it does not imply that any of the Führer's daydreams or Himmler's rather more precise thoughts would ever become concretised. Indoctrination was easy, but a survey of continued rural migration even in west Germany shows what an uphill task the party faced. Backe virtually admitted this in November 1942 in Posen. He expanded on the need to win the coming generation and argued that all that potential settlers needed was strong fists, a 'German heart' and a varied and thorough vocational education. As there was plenty of land, this was probably true. But then he came down to earth with the admission that in the previous year only 6,000 apprentices had secured a certificate in agricultural-domestic science studies; he described the total as a drop in the ocean *(ein Tropfen auf einen heissen Stein).* Consequently he ordered the sixty to eighty best *Erbhöfe* in each *Kreisbauernschaft* (more or less equivalent to one in each village) to take on by 1 April 1943 a male or female apprentice each.[75]

Clearly, youth still had relatively little interest in the land, otherwise such measures would have been unnecessary. As one contemporary admitted, parental opposition was still strong. Besides, not all peasants were keen to take on young people and train them anyway. This had been said of them before the war, but once cheap Russian labour was available, their reluctance was sometimes intensified.[76] Here was another vicious circle in the Third Reich; to settle the east and replace foreign labour youth had to be trained, but the very existence of the 'white coolies' from the east worked against this in the first place. Backe's directive was an attempt to break this impasse but that it had to be issued at all is informative.

Of course, there were some possible recruits among adults. To begin with, many rurally-born Germans disinherited under the provisions of the EHG had gone to the cities prewar. It seems reasonable to assume that at least some might not have done so had a holding been made available in the east. Doubtless this would also have applied to absconding farm labourers. Further there were the *Volksdeutsche* from outside the 1939 frontiers, of whom some 400,000 were re-settled in the annexed eastern provinces. The military or para-military organisations were a particularly promising recruiting-ground, in view of the *Wehrbauer* concept, hence the SS rhapsody that 'The German east waits for the returning soldier and for the *Wehrbauer* of the SS'. A long-service soldier who went on the land received a gratuity of RM 16,200 instead of the normal RM 8,000.[77] The SA was another possible source; a directive of December 1941 called on its leaders to carry out a drive in the ranks by informing all members of what opportunities awaited on the land, and try to encourage them to acquire a settler's certificate.[78] Even other Germanic peoples were brought in. An 'Eastern Company' was formed in the Netherlands to stimulate Dutch interest, but little was achieved.[79]

It does not seem very likely that all these streams when joined would have exactly resulted in a flood particularly as settler selection remained as rigorous as ever, despite Backe's hint of relaxation of the criteria (see page 245). Biological health, character and political opinions all had to remain unobjectionable.[80] The statement, 'Candidates who have been married for a long time without having had children are unsuitable' showed that priority still lay with the concept of the peasantry as the life-source of the nation. If this was always to be the case it is hard to see how any real mass transfers could ever have taken place, especially as Himmler was determined to base his plans on the peasantry anyway. Hitler's talk of ten or twenty million Germans going east does not sound like a viable proposition.

Darré's downfall forms the postscript to the foregoing account of wartime events. Named as a Minister in 1933, nine years later he was virtually in disgrace and no longer effectively in power. This is a surprising end for the man who did so well in organising the delivery of the peasant vote for the party, no mean service. It is true that he was no 'Old Warrior' of the movement, and was never one of Hitler's intimates.[81] Nevertheless in 1933 he had been described as one of the 'most gifted young men in the circle around Hitler'.[82] Then began the long decline, beginning with the poor harvests and food crisis of 1934-35. His prestige suffered a further blow with the row over Meinberg in early 1937, when he stated that only through a 'superhuman effort' had he been able to win back the Führer's confidence. Finally, his persistent importuning of Hitler on behalf of his beloved peasantry probably rather got on his leader's nerves. Irritation on the one side accompanied disillusion on the other. In September 1939 Darré apparently tried to resign but his offer was rejected probably in order to preserve the façade of unity so dear to the Third Reich. In spurning the resignation the Führer gave his opinion of the whole RNS in words quoted at Darré's postwar trial, 'Your dirty agrarian-political apparatus isn't worth powder and shot. What have you actually achieved?'[83]

Coincidental with all this was the rise to prominence of Herbert Backe. Nominated as Food Commissioner for the Four Year Plan he began to emerge as a personality in his own right. Hitler had clearly ceased to have much confidence in Darré by 1939, and two years later he and Goering by-passed him completely regarding the organisation of the foodstuffs sector for the war against Russia. Instructions were given to Backe directly, under orders to keep them a secret from Darré.[84] A year later Goebbels was confiding to his diary that the Food Minister was completely and deservedly discredited with the Führer. A few days later he recorded Hitler's decision to get rid of Darré, and replace him by Backe. 'He [Backe] is not a pale theoretician . . . but a real first-class practical man.'[85]

This was one long-standing and perhaps decisive complaint against Darré. He had simply overdone the 'Blood and Soil' fuss over the peasantry and failed to get better results in the *Erzeugungsschlacht.* Goebbels' use of the word 'practical' for Backe sums up the difference between them. Virtually everyone who knew the RBF agreed he was no man of action, whereas his successor was less dreamy and more business-like. Werner Willikens declared at a postwar interrogation that it was always hard to get a decision from Darré, which led to general delays in Ministry business. Another of his confidants said that 'he had never baked bread in his own oven'. His eventual replacement by Backe was to some extent merely a question of personal administrative failure. Above all, he was apparently of little use in wartime: it has even been suggested that he paid little heed to the question of managing the food economy at all.[86]

In the end his interminable accentuation on 'Blood and Soil' began to get on the nerves even of some party comrades. Of course, ultimately all National Socialists subscribed to the concept of race, but agreement in principle marked variations in practice. Hitler undoubtedly believed in the theory of eugenics, as well as being a rabid anti-semite: these were perhaps the only two things in which he ever did genuinely believe, apart from Social Darwinism. Everything else was opportunism, including his careful cultivation of the rural vote after 1929. But in the anthropological side of race such as skull measurement or the identifying of blood groups he took apparently little interest. In speeches and conversations he seems to have used 'Aryan' virtually as synonomous with 'ruling-class'. The investigations of the 'Blood and

Soil' fanatics into whether individual peasants were Nordic or Alpine was decidedly a fringe activity in the Third Reich, except in the case of Himmler. Whether Hitler, Goering and Goebbels ever took all this seriously is at least open to question. If the Führer ever thought of it at all he probably considered it simply as a useful propaganda element to increase German nationalism, his own version of the Platonic myth about the men of gold, etc. As to his well-known Social Darwinism, history as a struggle between peoples, doubtless the Führer subscribed to this in all seriousness, but as an idea this has no necessary connection with the romanticism about Nordics. It could be joined in practice to both anti-semitism and eugenics, but it did not require any mysticism about blue eyes and long skulls.

Opposed to Hitlerian pragmatism was the true romanticism of men such as Darré, Himmler and Rosenberg, with all the fuss about the peasantry and its role in German history, the revival of folklore, etc. For Hitler the agrarian community was important for practical reasons; it impeded Marxism and yielded food and military material. As those considerations began to take precedence over 'Blood and Soil' from 1937 onwards, Darré began to lose what influence he had ever possessed. Once the shooting-war started the men Hitler needed were Backe and Himmler; they were both organisationally gifted and in no way opposed to exploitation of 'inferior' peoples. Hitler's decision to put the eastern areas under Himmler, even their settlement projects, was a sign of the times in 1939. By this time the *Wehrbauer* concept was beginning to take precedence over the old settlement programme; as Backe had been advocating something similar since 1925 (see page 153) he found it easy to work with Himmler. It may well be that this fact was as important ultimately for his rise and Darré's downfall as any difference in personal qualities. The latter's quarrel with the Reichsführer SS in February 1938 is therefore an additional factor to be weighed. That, and Darré's incompetence, effectively put an end to 'Blood and Soil', or rather, to Hitler's tolerance of it as an end in itself.

CONCLUSION

This final chapter has three aims: first, to evaluate National Socialist agrarian measures in their immediate context, in other words, how effective were they in view of Germany's actual economic and political situation? Secondly, their place in the general framework of German history has to be considered, and lastly, what light can they throw upon Hitler's foreign policy?

In January 1933 the economic blizzard then raging over Germany was already beginning to abate. None the less, throughout the next eight years the position of the country was delicate, faced as it was with a perpetual balance of payments crisis. So National Socialist policy had two sides, one ideological and the other more immediately practical. The first was the preservation of the peasantry as a bulwark against the modern world of liberalism and socialism; the other was the desperate need to maximise domestic food production. In this way, currency could be saved, and utilised to buy other raw materials, such as those required for rearmament. Clearly, in practice, there was often little difference between the two aspects since increased output, and therefore bigger farm incomes, would itself assist in the conservation of the peasants as a class. However necessary these latter aims were, however, the party could never see them as sufficient in themselves; even with higher prices the farmer could still come unstuck under a free market system.

Such considerations led to the core of what was more purely National Socialist, the introduction of security of tenure for the peasant. In this the new régime was substantially assisted by previous legislation. To a very real extent, the German grain-producer had become cut off from the vagaries of the world market by 1933. National Socialist ordinances extended this isolation to its logical conclusion by affording guaranteed prices for the full range of agricultural products. This was complemented by the *Erbhof* and settlement laws, forming an internally-coherent programme to rebuild the peasantry. If the initial value-judgements about the landed population are accepted, then the government's policies were perfectly valid and rational.

Where they were less successful was in the maximisation of output. Some increases were obtained, but for various reasons the drive to produce more food could have achieved even better results. By 1938 migration from the land was undoubtedly the biggest single factor here. This in itself relates to Hitler's reluctance to give higher prices to farmers, due to his fear of inflation and his need to keep German exports competitive. So food costs had to be kept stable for workers. Hence the constant guerilla war between the RNS and Hitler in 1938-39, leading to the bitter observation of one report that the Führer's promise to restore profitability to agriculture had never really been maintained.[1] In late 1942 Backe was still talking about the need to tackle the problem of farm work's relatively low rate of remuneration, just as the RNS had done before 1933. The 'price-scissors', the term for the gap between industrial and agrarian prices, was never completely closed. So as far as the advocates of a 'peasant policy' were concerned, the time-span 1933-36 represented a false dawn. It really looked then as though farmers were going to be singled out for preferential financial treatment, as their incomes rose more rapidly than did that of other sectors. By 1937, however, the situation changed as unemployment diminished, and agriculture's share of the national income correspondingly declined.

The culmination of Darré's resentment at this trend was the memorandum of January 1939, where he flatly demanded higher prices to be cushioned by subsidies to the consumer. He did not get very far with Hitler, as usual, which gives rise to two comments. First, it was characteristic of the Führer to disregard policy details; soon after his arrival in office he told Lammers that he did not wish to be bothered with routine documents.[2] In respect of the agrarian sector, he certainly adhered to this: Darré never found it easy to get his ear. On the other hand, he might have complained with justice about the persistent importuning by the agricultural lobby. After all, Darré had accepted the need for German rearmament. Logically he could hardly deny afterwards the consequences which this entailed. In May 1935 he advocated stable wages for industry, so the corollary of equally stable living-costs could hardly be rejected. In any case, as agrarian incomes went up between 1933 and 1936, Hitler doubtless felt that he had met his moral obligation to the farming community. Moreover, the peasants themselves, although obviously generalisations here are potentially misleading, stood behind the party to a very real extent.[3] One postwar report has even suggested that those in Schleswig-Holstein could look back on the era with no particular ill-feelings.[4] Doubtless the judicious flattery in which the party indulged on the land played a role here, as well as the material advantages already mentioned. In any case, Hitler enjoyed widespread electoral support from the rural population prior to 1933, so the NSDAP started out with a considerable fund of goodwill from those who tended to see it very much as their government.

It was sometimes unavoidably the case that either individual farmers or an entire community would show dissatisfaction with some aspects of governmental policy. There was some boycotting of the Harvest Festival in the Aachen area in 1934, for example, and on another occasion a whole village refused contributions to Winter Help.[5] The Habbes affair caused a certain reaction in Westphalia. In Munich there was even a public demonstration in 1938 as a result of labour shortages.[6] But such instances can hardly be called serious resistance to the régime. Disappointment at being left to manage a farm with inadequate help does not imply a wish to change the government, but almost certainly merely that it should modify its policy in a certain respect.

This applies also to the reception of anti-Christian propaganda in some areas. A special RNS calendar for peasants had to be withdrawn from the Rhineland in 1935 as its tone offended local religious susceptibilities.[7] Equally, some folklore cults were not always well received, giving rise in one instance to a reference to the RNS as 'the new heathens'.[8] Darré was well aware that here was an area of potential opposition, where the party had to tread with immense care. The whole ethos of blood and race was bound to clash with religious convictions to some extent, hence his own view that 'the foundations of the Christian doctrine should be pulled away without actually mentioning Christianity'.[9] In some cases, a less cautious approach gave rise to trouble, but again this should not be exaggerated. Disgust over racial propaganda does not necessarily entail the belief that the NSDAP's policies should be rejected root and branch.

The general acceptance of the party on the land clearly has some relevance to its relative mildness in respect of agrarian administration (which can hardly in itself have failed to add to its popularity). The peasantry got the carrot rather than the stick. To those familiar with the concept of the Third Reich as a police state without equal this may seem strange. But without a high degree of self-sufficiency the whole economic recovery programme would have collapsed, so that the NSDAP was bound to be careful how it dealt with the peasantry, for purely practical reasons. As Darré pointed out, a gendarme cannot supervise the right use of a manure-heap.[10] Moreover, regulations did not always work in practice anyway, as in the case of the wartime Black Market; nor did they impede rural migration prior to 1939. To some extent, this was due to deliberate obstructionism by members of the agrarian administration who were not supporters of the régime. Delays in obtaining land for settlement in 1934 was attributed to them in a couple of areas.[11] The following year Darré complained about how hard it was for him to get some of his orders executed.[12] So a whole complex of factors, political and administrative, conspired to mitigate the effects of some party measures. In the case of anti-religious propaganda, public opinion on the land was itself probably the greatest single influence in this respect.

A further instance was the *Erbhofgesetz* which was often applied less rigorously than it might have been. Again, a cautious approach was adopted, for fear of upsetting the peasantry. Whether this concessionary attitude would have been maintained after the so-called transitional period is obviously doubtful.

This leads to the point that any final assessment of National Socialist agrarian policy has to be based on the simple premiss that Hitler did not actually intend to lose the war. In other words, what would have been the long-term effects of his measures? Clearly, a German victory would have enabled resources to be switched to agriculture from rearmament. Hence Darré's talk about a gigantic postwar project for rural transformation (*Aufrüstung des Dorfes*).[13] Backe committed himself to talk of a ten-year programme, costing RM 66,000 million.[14] Whether this would ever have been implemented cannot now be ascertained, but there was no financial reason why not, as it represented roughly what was spent on armaments between 1933 and 1939, at a time when on average Germany had a lower GNP than in the latter year.[15] Moreover, the fact that Hitler refused to sanction more funds for agriculture before the war is no guarantee that he would not have done so afterwards. All this is, however, based on the concept of a German victory, and ignores the more important issue, namely, whether the

NSDAP's policy could have worked without a war.

This seems much less likely, since by 1933 Germany was already so industrialised, and city life so much more attractive than that on the land. As one farmer's wife put it 'nowadays they [peasants' daughters] would rather wear silk stockings than clogs'.[16] People did not always leave the villages on purely financial grounds, but rather because the monotony and the long hours acted as a deterrent, especially on the younger generation. It appeared to be the case that the majority of people were perfectly content in industrial western Germany, and it is hard to believe that even the most radical programme of rural development would have changed this to any really significant degree. This was equally true of farm labourers. Without a war there would have been little chance of them rising in status to independent farmers, as so little land was available. In fact, state aid was increased in 1939, but the impact could only have been slight.[17]

Continuing labour losses would have put the farmers (and their wives) under ever more pressure. A greater degree of mechanisation would have afforded some compensation, it is true, and there is no doubt that the régime was aware of this. It formed the focal point of Darré's postwar project already referred to. But the insistence on the peasant holding was bound to put limits on technical progress in the long run, as postwar west German governments know to their cost. In retrospect, it would appear that the NSDAP really arrived on the scene too late to reverse the train of German industrial development.

These considerations obviously relate to the remarks made earlier about the rationality of Hitler's agricultural policies. In the final analysis, however well worked out were the means for achieving the party's ends, it seems unlikely that the normal course of German socio-economic development could have been halted for very long; the programme of 1933 was doomed to eventual failure.

No area illustrates this more clearly than that of settlement in the east. Whatever the régime said about living-space, in 1938 the main territorial issue in the Third Reich was the steady depopulation of the border areas, despite an internal colonisation programme, and official exhortations to 'Go east, young man'. It is here that the advocates of 'Blood and Soil' appear most sharply divorced from reality. True, occasionally some recognition of fact broke through, as when Backe wrote in 1933 that whereas the Germans were a people without space, their own east was a space without people.[18] Germans had a reluctance to resettle, which military conquest did not seem to change. A waiting-list for farms in the Ukraine maintained by Erich Koch contained only 237 applicants in 1943.[19] Despite Himmler's willingness to remove the original inhabitants, few Germans seemed to want to move in. In this respect even a German victory would have been Dead Sea fruit, in providing the Third Reich with new land when its citizens were busy depopulating part of their existing countryside. For the party it was, therefore, not a valid excuse for the relative failure of the prewar settlement scheme to say that land shortages had inhibited it. What light this casts on Hitler's foreign policy will be discussed presently.

The second part of this Conclusion looks at NSDAP measures in the wider context of modern German history. Nothing is more striking than the continual recurrence of so many themes in trade and agrarian concepts from the nineteenth century onwards, that is, roughly from the onset of industrialism. This explains why so much party propaganda sounds like a simple repetition of earlier statements from other bodies. In Germany, as in some other Continental countries, the dividing-line came to be the attitude to free trade, because it was linked with political liberalism, and could not be seen as a purely economic measure. So debates over import duties, whether in the 1870s or in the 1890s, at the time of the 'Caprivi crisis', tended to become controversies distinguishing two wholly different outlooks from one another. Protection and self-sufficiency were seen as the best guarantors of the existing way of life for the eastern squirearchy since it ensured for them a market for their grain by excluding foreign competition. The writing was on the wall for the Junkers by the 1890s, as Germany was becoming ever more an exporting country, and the alliance between them and the western industrialists began to weaken.[20] After all, an accord with heavy industry, 1879 style, was one thing, since German coal and iron did not go abroad. But exports did, and the people who sold them could not afford any retaliation against high German customs duties, so that they were naturally in favour of low tariffs (and cheap foreign food for their workers).

Even the advance of industrialisation in itself seemed to carry a potential threat to the settled way of life in Germany from which the Junkers were the chief beneficiaries. Rural migration from east of the Elbe to western urban conurbations left estate-owners short of labour, and gradually Polish workers filled the gap. To the dismayed land-owners, German areas took on an ever greater Polish air, whilst the

German erstwhile farm labourers became an urban proletariat, and frequently fell victims to Marxism. Hence, the settlement projects of the 1880s already discussed (page 141) and the foundation in 1893 of that classic pressure-group, the League of Farmers (the original title of the *Landbund*).

It is interesting to look at some of the League's concepts in the light of the later rise of the NSDAP. For instance, one of its spokesmen, von Kanitz, proposed a government grain monopoly with fixed prices, in 1894. A contemporary publicist, Gustav Ruhland, employed by the League, elaborated the idea into a whole system of agrarian economics. His works were published as a trilogy in 1912 and then disinterred by Darré to serve as a model for the RNS structures.[21] Ruhland's importance should not be over-emphasised, but his concept of 'German (i.e. national) Socialism' as opposed to the Marxist variety survived to be landed in the 'Blood and Soil' publications of Hitler's Germany.[22]

The League illustrates in its advocacy of security of farm tenure in the twenties a further side to eventual National Socialist policy. The intellectual background of the *Erbhofgesetz* has been mentioned in the text, and it suffices here merely to reiterate that the Act of 1933 embodied much that was earlier, non-party thinking. Similarly, National Socialist attacks on the 'Jewish-materialist' Civil Code of 1900 which permitted the fragmentation of family farms, paralleled earlier propaganda from Right-Wing groups.

There was another aspect to the demands for Protectionism which also deserves mention, especially as it emanated from writers who could hardly be called pioneers of National Socialism and which virtually coincided with League agitation. A whole school of thought, centring around members of the celebrated *Verein für Sozialpolitik* began around the turn of the century to cast doubts upon Germany's future status as a wholly industrial nation. Chief among them was Adolf Wagner, whose gloom was connected with the country's diminishing degree of self-sufficiency in food.[23] Once Germany became dependent on others for its nourishment, they might exploit the situation by refusing anything but low prices for German industrial goods. Moreover, when Wagner (and Max Weber) surveyed the world trend towards huge, autarkic empires (the USA, Britain, Tsarist Russia) they asked themselves where their country could find sales outlets in future. Circumstances seemed to be pushing Germany towards a policy of self-sufficiency, or to a Central European customs union.

Such thinking resulted in a Conference in 1902 called by the Verein to discuss the whole issue of Germany's economic future. Here the respected agrarian writer Max Seering (later a critic of the EHG) brought out a programme for agriculture strikingly similar to actual post-1933 measures. Comprised in the package were farm inheritance reform (less drastically than in the EHG), grain duties, eastern settlement and the end of foreign labour on the land. The purpose of these observations is not to suggest that the named writers were National Socialists, but rather that the later programme of the NSDAP fell on well-prepared ground.

The events of 1914-18 did nothing to diminish the concept of the need for autarky in Germany, particularly in view of the Allied blockade, and the loss of Russian imports. Darré himself said that in building up the RNS he had always had the 1918 famine in mind. As many articles in the Third Reich put it, he who holds the bread bin of a nation in its hand can dictate its foreign policy: in other words, no autarchy without autarky. German attention was focused on to Central Europe as a self-contained economic unit by *Mitteleuropa*, Friedrich Naumann's work in 1915, which proved enormously popular. Then, as stated previously, the protectionist policies pursued by Germany's neighbours post-1930 began to push the Weimar Republic to high duties in retaliation. National Socialist vaunting of the domestic market as a cure for national economic ills was in effect a combination of previous notions about self-sufficiency in theory and a response to an immediate given situation.

Connected with this, and with settlement plans, was the idea of living-space. Again, this was hardly novel; as early as 1894 the Pan-German League was airing its concept of 'elbow room' for Germans in Europe.[24] As only advanced nations had a right to existence 'such inferior peoples as Czechs, Slovaks and Slovenes' would have to yield. 'The old drive to the east should be revived', to build a defensive wall against the Slavs. To make this dream a reality, a man must arise; a despot, since his policy would frighten the bourgeoisie. Three years later the actual expression 'living-space' (*Lebensraum*) was used, by a zoologist Friedrich Ratzel, an exponent of Social Darwinism.[25] His approach was essentially that 'struggle for existence' meant finally 'struggle for space'. It is informative that a copy of his work *Politische Geographie* was available to Hitler in the library of Landsberg fortress in 1924; although it cannot be said with certainty that the future

Führer ever read it, but the possibility is strong. Incidentally, Ratzel never actually advocated military conquest as a solution for Germany.

As with autarky, the war reinforced the idea of living-space, as Germany lost 14 percent of her territory and her overseas colonies. Hence the enormous popularity of the 1926 novel *People without Space* (*Volk ohne Raum*), which sold over a quarter of a million copies in seven years.[26] German victories in the east had already directed attention to its possibility as a source of land. The agitation in 1915-16 by agrarian associations for annexations there are now well documented.[27]

So several prewar agrarian policy ideas survived, sometimes in a sharpened form, the defeat of 1918. When, in 1929, the capitalist economic order began to break down, it seemed to a movement which had inherited so many of these beliefs that its hour had finally struck. As Backe wrote, 'We are at a turning-point, at the collapse of liberalism and the breakthrough of the racial idea.'[28] Liberalism had been under fire, especially in its aspect of ever-developing free trade between nations, for decades from the German Right. To this, the NSDAP added a second point that laissez-faire indifference to social problems ineluctably bred Marxism.[29]

The party was therefore to a very real extent in the mainstream of much previous thought in Germany, most of which developed in the nineteenth century. Its general indebtedness to the past can easily be illustrated by reference to the Romantic School, and some of its outpourings after 1800. As one agrarian historian has written, these writers 'gave priority to the whole over the individual, to the organic over the rational, to vocationally-based structure (*ständische Gliederung*) over market forces'.[30] These attitudes all found exemplification in practice in NSDAP measures. The *Erbhofgesetz* is an instance of the first tendency listed above, the myth of 'Blood and Soil' of the second, and the RNS of the third. It is therefore in no way surprising that the views on the peasantry of one member of the school, Wilhelm Riehl, could be quoted by Darré in his own first book in 1929.

The fact is that most NSDAP ideas originated in the nineteenth century; to the complex given above may be added Social Darwinism, anti-liberalism and racialism as a deliberate quasi-scientific programme. Since liberalism was believed to be the progenitor of Marxism, the inclusion of the latter as another 'anti' aspect of the party's programme in the twenties was a development of earlier thinking, rather than a departure. In the final analysis there was little, if anything, in the NSDAP's policies that was original, despite the following claim: 'The revolutionary [side] of the National Socialist outlook lies undoubtedly in the fact that for the first time in history we have recognised the laws of nature as being valid also for mankind.'[31] Many had, in fact, done so before, but unlike the NSDAP they did not acquire office. In effect, the reason why the party's agrarian programme was unworkable in the long run was because circumstances conspired in 1933 to bring a nineteenth-century movement into power in the twentieth.

However, the indebtedness to previous thinking should not be exaggerated, for three reasons. First, Hitler understood very well that negative propaganda against his adversaries was not enough, positive measures would have to be introduced to combat Marxism and liberalism. As he wrote in *Mein Kampf*, only a child would believe that a worker seduced from Communism would automatically enter a bourgeois party. Hence the campaign against bad employers on the land for their anti-social attitude, the advocacy of the Heuerling status for farmworkers, and the (admittedly limited) attempts to enable labourers to become settlers in their own right. It was in a sense a kind of swap. In exchange for free, collective bargaining and full political rights, the worker acquired a degree of social welfare and a share in Germany's enhanced status. In this respect the prefix 'National Socialist' tells its own story, as the adjective precedes the noun, whereas to an orthodox Marxist being a worker takes priority over nationality. According to Hitler at the 1933 Harvest Festival, liberalism stresses the individual, Marxism the social class, whereas National Socialism concentrates on the people, i.e. as an entity in itself. So the aim of the NSDAP's land programme was simply to overcome the class-struggle between peasant and labourer by integrating them both within the national community, and placing their common ethnic origins in the forefront of their thoughts. However little genuine value can be given to 'Socialist' in the party's title, it was a populist movement, unlike the DNVP or previous nationalist groupings, and as such could preach a social doctrine and make it sound convincing, in a way that Right-Wing predecessors could not. Although indebted to them for some of its ideas, the NSDAP was far more than just their heir.

Secondly, despite the neo-medieval trappings of some areas of Hitler's Germany, much that happened was fully in the line with modern developments. For example, guaranteed prices for farm products came to be an accepted solution in many European countries, irrespective of their régime. The Popular Front in France introduced a fixed grain price in 1936. Even countries with long liberal traditions, such as Britain, brought in state or quasi-state structures to aid agriculture in the thirties, the Milk Marketing Board being one illustration. This century has been marked by a general growth of the public sector, and restrictions on laissez-faire; the NSDAP's attitude to private farm property, namely, that ownership must be linked to a sense of social responsibility, probably sounds less strange to Western European ears now than when it was first formulated by Werner Willikens in 1930. Similar observations might be made about the use of public money in the Third Reich to stimulate employment, or the incomes and prices policy as laid down in 1935.[32] It is surely not too fanciful even to compare the imported landworkers of 1938-39 to the modern Gastarbeiter, as after all they too were replacing Germans who had gone off to better jobs. In some respects quasi-medieval, in others fully in the context of Western development, Hitler's Germany presents a curious paradox.

Thirdly, in assessing to what extent agrarian policies of the NSDAP derived from the past, we have to remember that the 'Blood and Soil' group around Darré was to some extent isolated even within the party, let alone in contemporary society as a whole. Hence the constant friction described in Chapter 7 between its members and other bodies, including the DAF. In other words, we should beware of assuming that all National Socialists necessarily subscribed to the same views. The whole RNS organisation was highly unpopular in many quarters, as witness the uncomplimentary remarks often made of it by some sections of the NSDAP.[33] The party was really a coalition of everyone against the 'System' of parliamentary democracy rather than one founded on commonly-held economic principles. One member later testified that on joining he had not felt that he was in a normal movement at all, because of the blocks of former DNVP or even ex-Communists whom he encountered in the ranks.[34]

What attracted many to the party was its belief in action, on occasion virtually as an end in itself. This dynamism, often expressed in military terminology, was common both to the Third Reich and to Mussolini's Italy, with its Futurist links. Everything was a struggle; the National Socialist work-creation programme was the 'Battle of Work', the drive for more food the 'Battle of Production' (and in Italy the

'Battle of Grain'). It was in the nature of Fascism to present its peoples with gigantic, quasi-military tasks, perhaps partly to justify the dictatorship, and partly to submerge individual personality. As Rosenberg wrote, 'The style of the new Germany is that of a marching column, no matter where, or to what end.'[35] It does not by any means follow that everyone caught up in this necessarily shared the same views, which applies to the party as well as to the masses.

Finally, what light does agrarian policy throw on foreign affairs? It is obvious that a strong connection existed between internal and external matters after 1933; settlement and autarky are clearly bound in with the notion of living-space, since the first two could hardly be achieved without the latter. It has now become axiomatic, at least in some quarters, that the main driving-force behind Hitler's foreign policy was the desire for *Lebensraum* and raw materials. Certainly, it was one of his major themes in discussions with either visiting foreign statesmen or with his own generals or party members. These comments are now so well known that repetition would serve no useful purpose.

However, an examination of trade policy in the Third Reich shows an interesting development along rather different lines. The country was steadily improving its position over food supplies in a variety of ways. Domestic output had been significantly augmented, and still existing gaps in fats or fodder covered by bilateral treaties, either with Balkan countries or such neighbours as Denmark and the Netherlands, supplemented by an expansion of the whaling-fleet. Whether this was initially a consciously-planned programme or not is beside the point; here in any case Darré's own enthusiastic support in 1934 for such arrangements is informative, on the grounds that trade with the Netherlands would pull her into Germany's political orbit (see page 163). Backe was later to claim that the country had worked consciously towards the idea of a better-integrated Central European economic policy since 1933, citing the bilateral agreements in support.[36] Whether this was really true, or merely the usual politician's habit of claiming the purely fortuitous as evidence of his own foresight, is difficult to establish. But the fact remains that by February 1938, von Papen could speak of a future commonwealth in south-eastern Europe, consisting of states under German political, economic and cultural leadership; Balkan lands would exchange food and raw materials for manufactured goods from the Reich.[37] The whole suggested system smacks of a seventeenth-century motherland surrounded by client colonial economies.

Darré developed the concept of a New Order in the European trade structure in a speech to the party's Economic Committee in January 1939.[38] His own line of approach was to take the RNS's reorganisation of agrarian marketing as the basis for the reconstruction of Central Europe's economy in general. Backe followed his chief three years later, but with one significant difference.[39] Wartime conquests now made it possible to think of a reorganisation which would also comprise the Soviet Union: the starting-point was his own (rejected) doctoral dissertation for the university of Göttingen in 1925-26. Ten thousand copies of this study of the grain trade in the Tsarist era were distributed to his staff prior to the invasion of Russia.[40] The cornerstone of his thesis was that by industrialising the Soviet Union the Communists were contradicting the economic geography of Europe: the whole area should really be the Continent's granary. In passing, the allocation of the role of a source of raw materials to the Russians neatly complemented National Socialist racial propaganda, noticeably much sharper in wartime than before it. The naturally inferior should have a dependent economy.

Backe was able to link Russia with south-eastern Europe by proposing the latter area as a supplementary cereal-supplier: the Netherlands and Denmark would furnish the dairy products. In exchange, there would be industrial goods from the western European countries, especially Germany. Autarky would be raised from national to continental level, in effect, based on the idea of specialisation by individual countries (*arbeitsteilige Gemeinschaft*). Seen in a wider context, the whole thing sounds like a sort of European apartheid, since dependence on overseas countries would have been greatly lessened, if not completely removed. Contemporary Food Ministry Officials played with the same thoughts; as one said (significantly in a lecture in Sofia) if all Europe had a marketing structure like Germany's, no fodder from outside the continent would be needed, even if full autarky was not necessarily desirable.[41]

His role of Food Commissioner in the Four Year Plan enabled Backe to extend his ideas to actual practice, as that organisation was entrusted with the exploitation of the eastern areas. Goering himself was also of the opinion that the Russian economy should be treated as colonial, even in the long-term, as his directive of 8 November 1941 made clear.[42] The same view was expressed in the survey drawn up by his planning staff earlier in the year (which included an REM representative). Southern Russia was in future to draw its manufactured wares from the west, and send over its food surpluses; to ensure a sale for German industry the present supplier (the Russian forest belt region) would be eliminated as a competitor.[43] In other words, wartime acquisitions were to be used to provide in effect captive markets for German exports. Thus from peaceful reorganisation of the Continent in 1938-39, a whole New Order based primarily on politico-military hegemony had gradually evolved.

However profitable, or successful, these projects may ever have proved is irrelevant for the moment, the point is that they had nothing to do with settlement or living-space. Although the Führer was wont to muse frequently on the number of Germans to be moved east, ten or twenty millions or whatever figure struck his fancy, the actual policy in the east was in fact just old-fashioned colonial exploitation. Incidentally, if the Ukrainians were to buy industrial goods as outlined by Goering, surely that implied that they would stay in southern Russia? If so, where was the living-space for millions of Germans? This thought did not escape several officials at the time, who queried exactly what was planned for the original inhabitants.[44]

Similar gaps between theory and practice were visible in prewar foreign policy; an interesting document in this connection is Darré's memorandum of January 1939.[45] First, he drew attention to the fact that the occupation of the Sudetenland and Austria had actually hindered self-sufficiency, since both areas were less able to supply their own food needs than the rest of the country (Germany in the frontiers of 1937). Secondly, although supplies of grain were available on the world market, neighbouring countries did not have a fats surplus to offer Germany. Hence Darré's unequivocal statement in his letter that taking foreign territory would not assist the Third Reich agriculturally, as it did not possess what Germany wanted. This point was later emphasised by Backe, when he stressed that the occupation of Bohemia and Moravia in March 1939 had left Germany much as it had been. Supplies of sugar were improved, those of pigs and fodder worsened.[46]

Thus, although it is easy to trace a continuity of thought in Adolf Hitler from 1924 right up to the outbreak of war regarding the need for more land to cater for the food supply of an increasing population, in reality his deeds contradict it. Whatever his beliefs about autarky in general, in practice, at least initially, he occupied the wrong places, even in the case of Poland. Although the annexed Warthegau and the satellite

General-Gouvernement did supply the Third Reich with meat, grain and sugar, as far as fats were concerned Darré's predictions were borne out. The latter area sent some small quantities to Germany but the Warthegau actually received on balance 26,000 tons of fats from the Third Reich (due to shortages in eastern Upper Silesia).[47] Indeed, Hitler's apparent disregard of the real foodstuffs position may well have contributed to the disillusionment of Darré in 1939. Incidentally, the fact that both Austria and Czechoslovakia were agriculturally valueless was so well known that even General Beck pointed it out in a memorandum in November 1937.[48]

Neither did either offer Germans any living-space; here again there is an important deviation from Hitler's own words. In the now celebrated Hossbach minutes of 5 November 1937, he spoke of the possible annexation of both countries as providing foodstuffs for five to six million people, based on the compulsory emigration of three million of the original inhabitants (where to?). After the invasions had taken place nothing even remotely like this occurred. Hitler's speech to some of his generals on 3 February 1933 is another example. He mentioned settlement as the only way of solving unemployment in the long term. As we have seen, no effort actually was made in this respect, and the workless were eventually absorbed into industry anyway. Thus whatever the Führer said about settlement, he made little attempt to follow to its practical conclusion; either the speech was propaganda, or Hitler was a poor economic forecaster. In any event, the gap between pronouncements and actual policy remind us once again that it would be unwise to attach too much importance to all of Hitler's statements, as though each were a declaration of intent, for which careful plans were to be drawn up. Incidentally, this latter speech also included the suggestion that increased exports were one possible way out of Germany's current difficulties. Since this was what the east was eventually to be used for, according to the directives in 1941 already quoted, we might suggest that Hitler's long-term aim was to secure a political hegemony inside which Germany's economic future would also be secured. Whether this could best be achieved via direct settlement or via the acquisition of new, captive sales outlets for German industry was very likely an option that Hitler left open to the last minute, in accordance with his normal pragmatism.

At any rate, the boundaries of his Reich from September 1939 corresponded almost exactly to the 'Great German' concept of 1848 (yet another nineteenth-century legacy) hence the ironic description of Hitler as Austria's revenge for Königgrätz. Hence presumably also the fact that Slovakia (under Hungary in 1848) was not annexed, neither was Central Poland, although both offered living-space for Germans as much as the Ukraine in principle, if not to the same extent.

To sum up, it seems improbable that any mono-causal explanation can be made to fit the course of events in the Third Reich: this study has been concerned with agrarian policy as such and therefore the main conclusion has to be what light analysis of it can throw upon the broader stage of external affairs. Certainly it does not seem to suggest that the pursuit of living-space was by any means the sole pillar of Hitlerian policy. Moreover, Germans in the main displayed a marked distaste towards eastern settlement, even inside their own frontiers. As Hitler usually had a wary eye on the pressure-gauge of German public opinion it would appear at least doubtful that he intended a forcible colonisation by his countrymen in the event of a military victory. In any case, when the officials of the *Reichsnährstand* arrived in the Ukraine they launched a campaign to de-collectivise the land, not to facilitate German settlement but to return the soil to individual Ukrainian peasants under the slogan 'Land to the vigorous cultivator'.[49] Doubtless Himmler had other plans for the inhabitants but this only illustrates again the gulf between many individual members of the NSDAP, even at the summit, an additional hazard involved in analysing policies. What Lord Haldane said of Germany in 1912 was even truer of the Third Reich: 'When you mount to the peak to [sic] this highly-organised people you will find not only confusion but chaos.'[50]

APPENDIX: THE ERBHOF IN STATISTICS

Table A1. Erbhöfe as at mid-1938 by size of holding

Size (Ha)	No. of Erbhöfe	As % of total no. of Erbhöfe	Agricultural land comprised (Ha)	Agric. land comprised as % of total of all Erbhöfe land
Under 7½	20,067	2.9	134,470	0.9
7½-10	99,786	14.6	875,521	5.6
10-15	175,444	25.6	2,168,463	13.9
15-20	118,741	17.3	2,053,121	13.2
20-25	75,696	11.0	1,692,212	10.9
25-50	145,057	21.2	4,969,085	31.9
50-75	33,120	4.8	1,975,355	12.7
75-100	11,320	1.7	964,612	6.2
100-125	4,680	0.7	520,397	3.4
Over 125	1,086	0.2	208,637	1.3
Total	684,997	100.0	15,561,873	100.0

Source: Vierteljahrsheft zur Statistik des Deutschen Reiches 1939, Sec. II, p. 36.

Table A2.

Number of actual Erbhöfe mid-1938 in each size category of holding compared to total number of holdings in the same category, by region and province

Prussian province	Actual Erbhöfe		Total holdings in province/region*		% of total holdings enrolled as Erbhöfe	
	7½-20 Ha	20-125 Ha	7½-20 Ha	20-125 Ha	7½-20 Ha	20-125 Ha
E. Prussia	24,911	24,332	39,366	26,666	63.3	91.2
Berlin	76	24	171	44	44.4	54.5
Brandenburg	18,595	21,580	32,877	22,779	56.6	94.7
Pomerania	25,351	18,450	38,936	21,268	65.1	86.8
Silesia	34,979	15,382	49,768	16,461	70.3	93.4
Saxony	19,816	14,697	24,661	14,881	80.4	98.8
Schleswig-Holstein	12,714	17,125	14,870	16,824	85.5	101.8†
Hanover	33,600	32,989	39,372	30,084	85.3	109.7†
Westphalia	19,125	16,436	25,137	16,256	76.1	101.1†
Hessen-Nassau	12,375	3,874	16,710	4,360	74.1	88.9
Rhine Province	12,133	4,208	27,958	6,161	43.4	68.3
Hohenzollern	647	207	1,750	253	37.0	81.8
PRUSSIA	214,322	169,304	311,576	176,037	68.8	96.2

Reichsregion						
Bavaria	101,442	55,873	139,264	60,365	72.8	92.6
Saxony	21,181	10,645	22,717	10,672	93.2	99.7
Württemberg	16,366	7,325	29,640	7,832	55.2	93.5
Baden	5,365	3,602	14,201	4,141	37.8	87.0
Thuringia	11,910	5,428	15,856	5,787	75.1	93.8
Hessen	6,503	1,041	9,289	1,221	70.0	85.3
Hamburg	309	290	354	293	87.3	99.0
Mecklenburg	2,600	6,715	6,840	7,048	38.0	95.3
Oldenburg	7,649	5,075	7,481	4,110	102.2†	123.5†
Brunswick	3,049	2,120	3,138	1,864	97.2	113.7†
Bremen	144	252	173	217	83.2	116.1†
Anhalt	1,480	940	1,713	935	86.4	100.5†
Lippe	800	974	1,006	916	79.5	106.3†
Schaumburg-Lippe	518	252	565	179	91.7	140.8†
Saarland	333	37	1,483	115	22.5	32.2
GERMANY	393,971	269,873‡	565,296	281,732	69.7	95.8

Source: Ibid p. 38.

*Excludes all farms publicly owned by state authorities, etc.

†Figures in excess of 100 percent are explained by the fact that the number of total holdings taken in each category are those for 1933, at a time when some of equal size may have been on lease and were therefore not included: if after 1933 the lease expired and they reverted to the original owner for management they were enrolled as *Erbhöfe.*

‡The total number of *Erbhöfe* given in the first two columns together (663,844) does not equal that in Table A1 as the latter includes 21,153 *Erbhöfe* either below 7½ Ha in size or above 125 Ha.

Table A3. Preliminary litigation: the enrolment of the Erbhöfe as at 1 July 1935 and the statistics for objections to enrolment/non-enrolment

Regional Court	No. of Anerbengerichte attached	No. of potential Erbhöfe in area	No. so far enrolled	No. of objections to enrolment or non-enrolment	Objections sustained	Objections rejected
Bamberg	56	38,267	29,732	5,950*	2,612	1,828
Brunswick	22	6,522	5,207	575	220	169
Breslau	115	81,702	55,900	12,018	6,689	3,870
Celle†	106	85,993	71,491	7,759	4,529	2,249
Darmstadt	48	10,680	8,456	1,876	715	736
Dresden	104	40,361	33,404	3,102	1,338	1,225
Düsseldorf	35	13,875	8,612	1,454	662	417
Frankfurt-on-Main	31	5,475	3,306	1,633	861	494
Hamburg	6	1,911	1,492	308	161	134
Hamm	99	49,798	38,567	3,800	2,327	1,101
Jena	63	23,368	17,904	2,204	994	850
Kammergericht	99	64,978	47,231	9,815	6,404	2,684
Karlsruhe	46	19,413	9,727	2,830	1,122	957
Kassel	60	20,370	15,794	1,955	1,065	649
Kiel	57	37,483	29,392	4,363	2,601	1,408
Cologne	69	28,674	11,002	6,782	2,859	2,238
Königsberg	63	66,703	47,404	8,984	4,606	3,160
Marienwerder	23	18,918	14,178	2,312	1,369	805
Munich	96	108,226	88,363	8,909	4,051	2,779
Naumburg	105	50,822	39,437	5,391	3,185	1,614
Nuremberg	57	52,880	39,807	5,392	2,321	1,470

Oldenburg	19	15,257	12,907	1,806	1,004	503
Rostock	23	15,904	9,267	1,074	598	260
Stettin	57	51,682	36,061	6,040	3,079	2,050
Stuttgart	59	39,871	26,006	5,069	2,307	1,509
Zweibrücken	26	5,318	3,122	1,038	565	287
GERMANY	1,544	954,451	703,769	112,439	58,244	35,446

Source: Dr Hopp, 'Die Anlegung der Erbhöferolle nach dem Stand vom 1 Juli 1935', *Deutsche Justiz* No. 1324, BA R22/2183.
*Includes objections both from peasants and Bauernführer.
†This is the ordinary Regional Court, not the Chief Court for Prussia.

Table A4. Litigation for Germany at Regional Court level 1935-39

Court	Year 1935	1936	1937	1938	1939	No. of Erbhöfe in area	Total no. of cases 1935-39
Bamberg	374	523	281	198	124	28,238	1,500
Brunswick	45	63	37	52	18	5,353	215
Darmstadt	31	67	84	55	26	8,753	263
Dresden	80	430	418	410	400	33,434	1,738
Hamburg	–	5	7	7	4	1,079	23
Jena	33	116	150	84	87	18,043	470
Karlsruhe	35	69	87	94	44	9,467	329
Munich	207	880	746	765	715	89,149	3,313
Nuremberg	197	434	308	255	167	40,986	1,361
Oldenburg	100	188	196	230	116	13,372	830
Rostock	28	112	100	89	63	9,906	392
Stuttgart	93	211	138	138	153	24,593	733
Zweibrücken	21	49	45	17	13	3,303	145
Celle*	1,105	3,041	3,180	2,927	2,311	407,000	12,564
Cologne	53	54	2	–	–		109
	2,402	6,242	5,779	5,321	4,241	692,676	23,985

*This is the Court for all Prussia.

Source: Undated minute BA R22/2248. These are all cases referred to regional courts because of appeals against the original *Anerbengericht* verdicts.

Table A5. Litigation in Table A4 in respect of Celle (Prussia) broken down into Regional Courts cases, 1935-37

Regional Court	Year 1935	1936	1937	Total 1935-37
Berlin	120	331	320	771
Breslau	195	555	523	1,273
Celle*	120	442	576	1,138
Düsseldorf	20	57	67	144
Frankfurt-on-Main	30	54	37	121
Hamm	50	213	231	494
Kassel	10	88	93	191
Kiel	60	196	252	508
Cologne	130	174	90	394
Königsberg	165	356	398	919
Marienwerder	30	88	93	211
Naumburg	85	252	262	599
Stettin	100	235	238	573
	1,115†	3,041	3,180	7,336

*This is the court for the local area only.

†In Table A4 this figure is shown as 1,105, the discrepancy being part of the original records.

Source: Undated minute BA R22/2248.

BIBLIOGRAPHY

MAIN SECONDARY SOURCES

Abel, Wilhlem. *Agrarpolitik.* Vandenhoek & Ruffrecht, Göttingen 1967.

Backe, Herbert. *Das Ende des Liberalismus in der Wirtschaft.* RNS Verlag 1942.

Backe, Herbert. *Um die Nahrungsfreiheit Europas.* Leipzig 1942.

Bente, Hermann. *Landwirtschaft und Bauerntum.* Junker & Dünnhaupt Verlag, Berlin 1937.

Boberach, Heinz (Ed.). *Meldungen aus dem Reich.* Stuttgart 1966.

Born, Karl Erich (Ed.). *Moderne deutsche Wirtschaftgeschichte.* Kiepenheur & Witsch. Cologne 1966.

Boyens, Wilhelm F. *Die Geschichte der ländlichen Siedlung.* Landschriften Verlag, Berlin/Bonn 1959.

Bracher, Karl. D. *Die Auflösung der Weimarer Republik.* Villingen/Schwarzwald Ring Verlag 1964.

Bracher, Karl D. *The German Dictatorship.* Weidenfeld & Nicolson 1970.

Bracher, Karl D., Sauer, W. and Schultz, G. *Die Nationalsozialistische Machtergreifung.* Cologne 1960.

Brand, Marie B. F. von et al. *Die Frau in der deutschen Landwirtschaft.* Franz Vahlen Verlag, Berlin 1939. This is Band 3 of *Deutsche Agrarpolitik,* Eds Sering, Max and Dietze, C. von.

Brandt, Karl. *The Management of Germany's Agricultural Food Policies in World War II.* Vol. 2, Stanford UP 1953.

Braun, Otto. *Von Weimar zu Hitler.* Harmonia Norddeutsche Verlagsanstalt, Hamburg 1949.

Broszat, Martin. *Der Staat Hitlers.* Deutscher Taschenbuchverlag, Munich 1966.

Bry, Gerhard. *Wages in Germany 1871-1945.* OUP 1960.

Buchta, Bruno. *Die Junker und die Weimarer Republik.* VEB Verlag, Berlin 1959.

Carroll, Berenice. *Design for Total War.* Mouton, The Hague 1968.

Cecil, Robert. *The Myth of the Master Race; Alfred Rosenberg and Nazi Ideology.* B. T. Batsford Ltd 1972.

Clauss, Wolfgang (Ed.). *Der Bauer im Umbruch der Zeit.* RNS Verlag, Berlin 1935.

Dallin, Alexander. *German Rule in Russia 1941-1945.* London 1957.

Darré, Richard W. O. *Blut und Boden.* RNS Verlag, Berlin 1936.

Darré, Richard W. O. *Das Bauerntum als Lebensquell der nordischen Rasse.* Lehmanns Verlag, Berlin-Munich 1938.

Darré, Richard W. O. *Neuadel aus Blut und Boden.* Lehmanns Verlag, Berlin-Munich 1938.

Darré, Richard W. O. *Um Blut und Boden.* RNS Verlag, Berlin 1941.

Dellian, Edouard. *Es wächst ein neues Bauernrecht.* Berlin 1938.

Dietze, Constantin von. *Die gegenwärtige Agrarkrisis.* Junker & Dünnhaupt Verlag, Berlin 1930.

Dittmar, Hans. *Der Bauer im Grossdeutschen Reich.* Deutscher Verlag, Berlin 1940.

Drescher, Leo. *Agrarökonomik und Agrarsoziologie.* Gustav Fischer Verlag, Jena 1937.

Erbe, René. *Die NS Wirtschaftspolitik 1933-1939 im Lichte der modernen Theorie.* Polygraphischer Verlag, Zurich 1958.

Feder, Gottfried. *Der deutsche Staat auf nationaler und sozialer Grundlage.* Verlag Franz Eher Nachf., Munich 1933.

Feder, Gottfried. *Das Programm der NSDAP.* Verlag Franz Eher Nachf., Munich 1933.

Fischer, F. *Griff nach der Weltmacht.* Droste Verlag, Düsseldorf 1967.

Franz, Heinz. *Der Mensch in der Siedlung.* Mühlacker 1937.

Franz-Willing, Georg. *Die Hitler Bewegung.* R. V. Deckers Verlag, Hamburg 1962.

Frick, Wilhelm (Ed.). *Die Nationalsozialisten im Reichstag 1924-31.* NS Bibliothek Heft 37, 1934.

Gauch, Hermann. *Die germanische Odal – oder Allodverfassung.* Blut und Boden Verlag, Berlin 1934.

Gerschenkron, Alexander. *Bread and democracy in Germany.* Howard Fertig, New York 1966.

Goerlitz, Walter. *Die Junker: Adel und Bauer im deutschen Osten.* C. A. Starke Verlag, Limburg 1964.

Göhring, Martin. *Alles oder Nichts,* Vol. I. J. C. B. Mohr (Paul Siebeck), Tübingen 1966.

Grosser, Alfred. *Germany in Our Time.* Pall Mall Press, London 1971.

Guillebaud, Claude. *The Economic Recovery of Germany 1933-1938.* Macmillan, London 1939.

Guillebaud, Claude. *The Social Policy of Nazi Germany.* CUP 1941.

Haushofer, Heinz. *Die deutsche Landwirtschaft im technischen Zeitalter.* Verlag Eugen Ulmer, Stuttgart 1963.

Haushofer, Heinz. *Ideengeschichte der Agrarwirtschaft und Agrarpolitik etc,* Vol. 2. Bayerischer Landwirtschaftsverlag, Munich 1958.

Haushofer, Heinz, and Recke, Hans Joachim. *50 Jahre: Reichsernährungsministerium – Bundesernährungsministerium.* Bundesministerium für Ernährung, Landwirtschaft und Forsten, Bonn 1966.

Heberle, Rudolf. *Landbevölkerung und Nationalsozialismus.* Deutsche Verlags Anstalt, Stuttgart 1963.

Heiden, Konrad. *Geburt des Dritten Reiches.* Europa Verlag, Zürich 1934.

Hellermann, Friedrich C. von. *Landmaschinen gegen Landflucht.* Junker & Dünnhaupt Verlag, Berlin 1939.

Hertz-Eichenrode, Dieter. *Landwirtschaft und Politik in Ost Preussen 1918-1930.* West Deutscher Verlag, Köln/Opladen 1969.

Hitler, Adolf. *Hitler's Table Talk.* Weidenfeld & Nicolson 1953.

Hitler, Adolf. *Mein Kampf.* Franz Eher Verlag, Munich 1933.

Holt, John B. *German Agricultural Policy 1918-1934.* Univ. of North Carolina Press 1936.

Horkenbach, Cuno (Ed.). *Das Deutsche Reich bis heute.* Berlin 1930.

Jacobs, Ferdinand. *Von Schorlemer zur Grünen Front.* Hans Altenberg, Düsseldorf 1957.

Jasper, Gotthard (Ed.). *Von Weimar zu Hitler 1930-1933.* Kiepenheuer & Witsch, Berlin/Cologne 1968.

Klein, Burton H. *Germany's Economic Preparations for War.* Harvard University Press 1959.

Koehl, R. L. (Robert). *RKFDV: German resettlement and population policy 1939-1945.* Harvard University Press 1957.

Koeppens, Anne Marie. *Das deutsche Landfrauenbuch.* RNS Verlag, Berlin 1937.

Krausnick, Helmut et al. *The Anatomy of the SS State.* Collins, London 1968.

Kretschmar, Hans. *Deutsche Agrarprogramme der Nachkriegszeit.* Junker & Dünnhaupt Verlag, Berlin 1933.

Maser, Werner. *Die Frühgeschichte der N.S.D.A.P.: Hitlers Weg bis 1924.* Athenäeum Verlag, Frankfurt/Bonn 1965.

Mass, Konrad. *Der deutsche Bauer einst und jetzt.* Verlag Morris Diesterweg, Frankfurt-am-Main 1936.

Matthias, Erich and Morsey, Rudolf. *Das Ende der Parteien 1933.* Droste Verlag, Düsseldorf 1960.

Mayer, Jacob. *Max Weber and German Politics.* Faber & Faber, London 1956.

Meinhold, Willy. *Grundlagen der landwirtschaftlichen Marktordnung.* Paul Parey, Berlin 1937.

Meinhold, Willy. *Die landwirtschaftlichen Erzeugungsbedingungen im Kriege.* Jena 1941.

Meyer, Henry C. *Mitteleuropa in German Thought and Action.* Mouton, The Hague 1955.

Meyer, Konrad (Ed.). *Landvolk im Werden.* Berlin 1942.

Molitor, Erich (Ed.). *Deutsches Bauern- und Agrarrecht.* Quelle & Meyer Verlag, Leipzig 1936.

Mosse, George L. *The Crisis of German Ideology.* Weidenfeld & Nicolson 1966.

Muller, Josef. *Deutsches Bauerntum zwischen gestern und morgen.* Verlag der Universitatsdrückerei, Würzburg 1940.

Nicholls, A. and Matthias, E. (Eds). *German Democracy and the triumph of Hitler.* George Allen & Unwin Ltd 1971.

Noakes, Jeremy. *The Nazi Party in Lower Saxony 1921-1933.* OUP 1971.

Orlow, Dieter. *The History of the Nazi Party 1919-1933.* University of Pittsburg Press 1969.

Petzina, Dieter. *Autarkiepolitik im Dritten Reich.* Deutsche Verlags, Stuttgart 1968.

Pulzer, Peter G. J. *The rise of political anti-semitism in Germany and Austria.* John Wiley & Sons, New York 1964.

Rechenbach, Horst (Ed.). *Bauernschicksal ist Volkesschicksal.* RNS Verlag, Berlin 1935.

Reischle, Hermann and Saure, Wilhelm. *Der Reichsnährstand – Aufbau, Aufgaben und Bedeutung.* RNS Verlag, Berlin 1936.

Reischle, Hermann. *Die deutsche Ernährungswirtschaft.* Junker & Dünnhaupt Verlag, Berlin 1935.

Reischle, Hermann (Ed.). *Deutsche Agrarpolitik.* Berlin 1933.

Reischle, Hermann. *Kann man Deutschland aushungern?* Berlin 1940.

Reischle, Hermann. *Reichsbauernführer Darré, der Kämpfer um Blut und Boden.* Berlin 1933.

Rich, Norman. *Hitler's War Aims,* Vol. 1. Andre Deutsch 1974.

Rohr, von Georg. *Grossgrundbesitz im Umbruch der Zeit.* Georg Stilke Verlag, Berlin 1935.

Sachse, Karl. *Der Reichsnährstand.* Kohlhammer Verlag, Stuttgart 1941.

Saure, Wilhelm. *Reichsnährstandgesetze.* W. de Gruyter, Berlin-Leipzig 1935.

Schacht, Hjalmar. *Abrechnung mit Hitler.* Rohwohlt Verlag, Hamburg 1949.

Schacht, Hjalmar. *My First Seventy-Six Years.* Allen Wingate, London 1955.

Scheda, Karl (Ed.). *Deutsches Bauerntum – sein Werden und Aufstieg.* Verlag Karl Ehlers, Konstanz 1935.

Scheuble, Julius. *Hundert Jahre Staatliche Sozialpolitik 1839-1939.* W. Kohlhammer Verlag, Stuttgart 1957.

Schmitz, Hermann. *Die Bewirtschaftung der Nahrungsmittel und Verbrauchsgüter,* etc. Essen 1956.

Schneider, Hermann. *Unser täglich Brot.* Verlag Franz Eher, Munich 1930.

Schoenbaum, David. *Hitler's Social Revolution.* Weidenfeld & Nicolson 1966.

Schultzer, Joachim von. *Deutsche Siedlung.* Ferdinand Enke Verlag, Stuttgart 1937.

Schweitzer, Arthur. *Big Business in the Third Reich.* Bloomington USA 1964.

Simpson, Amos F. *Hjalmar Schacht in perspective.* Mouton, The Hague/Paris 1969.

Stolpher, Gustav, Hauser, Karl and Borchardt, Knut. *The German Economy: 1870 to the Present.* Harcourt, Brace & World Inc., New York 1967.

Stoltenberg, Gerd. *Politische Strömungen im Schleswig-holsteinischen Landvolk 1918-1933.* Droste Verlag, Düsseldorf 1962.

Taylor, Telford (Ed.). *Hitler's Secret Book.* Grove Press, New York 1961.

Thyssen, Thyge. *Bauer und Standesvertretung.* Karl Wachholtz Verlag, Neumünster 1958.

Tirell, Sara. *German Agrarian Policies after Bismarck's Fall.* Columbia University Press 1951.

Topf, Edwin. *Die Grüne Front.* Rowohlt Verlag, Berlin 1933.

Tracy, Michael. *Agriculture in Western Europe.* Fredk A. Praeger, New York 1964.

Treue, Wilhelm (Ed.). *Deutschland in der Weltwirtschaftskrise in Augenzeugen Berichten.* Karl Rauch Verlag, Düsseldorf 1967.

Vollmer, Bernhard. *Volksopposition im Polizei Staat,* Band II. Deutsche Verlags Anstalt, Stuttgart 1957.

Warriner, Doreen. *Economics of Peasant Farming.* Cass 1964.

Weiss, Leonhard. *Die Abmeierung.* Hans Buske Verlag, Leipzig 1936.

Wunderlich, Frieda. *Farm Labour in Germany 1810-1945.* Princeton University Press 1961.

JOURNAL ARTICLES, ETC.

Backhaus, Karl. 'Warum Malthus nicht recht behielt', *Der Vierjahresplan,* February 1944.

Bauer, Wilhelm and Dehen, Peter. 'Landwirtschaft und Volkseinkommen', *Vierteljahrsheft zur Wirtschaftsforschung,* Heft 4, 1938-39.

Behrens, Gustav. 'Stillstand in der Erzeugungsschlacht? ', *Odal,* Heft III, March 1939.

Beyer, Hans. 'Das Bauerntum Angelns während der grossen Krise 1927-32', *Jahrbuch des Angler Heimat Vereins,* 1961/2.

Beyer, Hans. 'Die Landvolkbewegung Schleswig-Holstein und Niedersachsens 1928-32', *Jahrbuch für Kreis Eckernförde,* 1957.

Beyer, Hans. 'Die Agrarkrise und das Ende der Weimarer Republik', *Zeitschrift für Agrargeschichte und Agrarsoziologie,* April 1965.

Blomeyer, Karl. 'Neuerungen im Erbhofrecht', *Jahrbücher für Nationalökonomie und Statistik,* Band 147, Gustav Fischer Verlag, JENA 1937.

Brandenstein, H. von. 'Die Auswertung landwirtschaftlicher Buchführungsergebnisse', Paul Parey Verlag, Berlin 1939 *Berichte über Landwirtschaft.*

Brugger, Peter. 'Der Anerbe und das Schicksal seiner Geschwister', *Berichte über Landwirtschaft,* Sonderheft 121, 1936.

Buchner, Hans. 'Ständischer Aufbau und wirtschaftliche Selbstverwaltung', *NS Monatsheft* 55, October 1934.

Clauss, Wolfgang. 'Bauer und Grossbetrieb im Kampf um die Nahrungsfreiheit', *Odal,* Heft 10, April 1935.

Darré, Richard W. 'Damaschke und der Marxismus', BA/NS 26/949, 1931.

Darré, Richard W. 'Die Frau im Reichsnährstand', *Odal,* Heft 9, 1933/4.

Darré, Richard W. 'Geburtenminderung durch das Erbhofgesetz? ', *Die Dorfkirche,* Heft 5, May 1934.

Darré, Richard W. 'Industrie und Reichsnährstand'. Speech to Reichsstandes der deutschen Industrie (at Krupp's invitation) 11.1.34: reprinted in *Odal,* April 1934.

Darré, Richard W. 'Ostelbier', Heft 12, *Odal,* July 1934.

Decken, von der, Hans. 'Die Mechanisierung in der Landwirtschaft', *Vierteljahrsheft zur Konjunkturforschung,* 1938/9.

Dickmann, Fritz. 'Die Regierungsbildung in Thüringen als Modell der Machtergreifung', *VJH* 1966.

Dietze, Constantin von. 'Bäuerliches Erbhofrecht und Bevölkerungspolitik', Heft I, *Odal,* July 1933.

Dietze, Constantin von. 'Deutsche Agrarpolitik seit Bismarck', *Zeitschrift für Agrargeschichte und Agrarsoziologie,* 1964.

Dommasck, H. 'Die Verbrauchsregelung in der deutschen Kriegsernährungswirtschaft', *Berichte über Landwirtschaft,* Vol. 28, 1943.

Emig, Dr. Karl. 'Das Reichsministerium für Ernährung und Landwirtschaft', *Das Dritte Reich im Aufbau,* Vol. IV.

Fensch, H. L. 'Die Unterbewertung der Landarbeit in den Betriebsgruppen der Landwirtschaft', *Berichte über Landwirtschaft,* Vol. 28, 1943.

Frauendorfer, Max. 'Deutsche Arbeitsfront und ständischer Aufbau', *NS Monatsheft* 54, September 1934.

Gies, H. 'NSDAP und landwirtschaftliche Organisation in der Endphase der Weimarer Republik', *VJH* 1967.

Horn, Wolfgang. 'Ein unbekannter Aufsatz Hitlers aus dem Frühjahr 1924', *VJH* 1968.

Jeurink, J. 'Berufswahl und Abwanderung der Landjugend im Kreise Göttingen', *Die Deutsche Berufserziehung,* etc., November 1942.

Kater, Michael H. 'Zur Soziographie der frühen NSDAP', *VJH*, April 1971.

Kinkelin, Wilhelm. 'Bauerntum und SS', *Odal,* Heft IV, 1936.

Koenig, A. 'Die landw. Berufs- und Fachschulen im Dienste des Ernährungswerk', *Die Deutsche Berufserziehung*, June 1942.

Kühne, Hans. 'Der Arbeitseinsatz im Vierjahresplan', *Jahrbücher für Nationalökonomie und Statistik,* Band 147, Gustav Fischer Verlag, JENA 1937.

Kühnl, Reinhard. 'Zur Programmatik der NS Linken: Das Strasser Programm von 1925/6', *VJH,* 1966.

Kummer, Kurt. 'Der Weg der deutschen Bauernsiedlung', *Odal,* Heft II, November 1939.

Kummer, Kurt. 'Neues Bauerntum im Dritten Reich', *NS Monatsheft,* Vol. 48, March 1934.

Lang, E. 'Produktion und Rentabilität der ländlichen Siedlung', *Berichte über Landwirtschaft,* Vol. 17, 1933.

Lange, H. 'Zur Frage der Berufslenkung unserer Landjugend', *Die Deutsche Berufserziehung,* April 1942.

Lange, Otto. 'Gustav Ruhlands System der politischen Ökonomie', *Odal,* Heft X, April 1935.

Loomis, Charles P. and Beegle, J. A. 'The Spread of German Nazism in rural areas', *American Sociological Review,* Vol. XI, December 1946.

Lovin, Clifford R. 'Agricultural reorganisation in the IIIrd Reich', *Agricultural History,* Vol. XLIII, No. 4, University of California Press.

Metzdorf, Hans Jürgen. 'Die Stellung der Kleinbetriebe in sechs Dorfgemeinschaften der Flämings', *Berichte über Landwirtschaft,* Paul Parey, Berlin 1939.

Müller, Albert. 'Soziale Arbeit in der HJ', *Soziale Praxis,* Heft 28, 1943.

Noakes, Jeremy. 'Conflict and development in the NSDAP 1924-7', *Journal of Contemporary History,* Vol. 1, No. 4. October 1966.

Petzina, Dieter. 'Germany and the great depression', Ibid., Vol. 4, No. 4, October 1969.

Preiss, Ludwig. 'Die Wirkung von Preisen und Preisänderungen auf die Produktion in der Landwirtschaft', *Berichte über Landwirtschaft,* Vol. XXIII, No. 4, Berlin 1938.

Proksch, Rudolf. 'Jugend aufs Land', *Odal,* Heft X, 1937.

Quante, Peter. 'Die Flucht aus der Landwirtschaft', *Odal,* Heft III, September 1933.
Rechenbach, H. 'Bauernschicksal ist Volkesschicksal', In Clauss, W. (Hrsg), *Der Bauer in Umbruch der Zeit,* RNS Verlag, Berlin 1935.
Reischle, Hermann. 'Agrarpolitischer Apparat und RNS', *NS Monatsheft,* Vol. 54, September 1934.
Reischle, Hermann. 'Die deutsche Ernährungswirtschaft', *Das Dritte Reich im Aufbau,* Band II.
Reischle, Hermann. 'Einkreisung der Unterwertung', *Odal,* Heft VIII, August 1939.
Ritthalter, Anton. 'Eine Etappe auf Hitlers Weg zur ungeteilten-Macht', *VJH,* 1960.
Scheda, Karl. 'Zur Agrartheorie Ruhlands', *Odal,* Heft I, July 1933.
Seraphim, Hans Jürgen. 'Neuschaffung Deutschen Bauerntums', *Zeitschrift für die gesamte Staatswissenschaft,* 1935.
Sohn, Frederick. 'Allgemeiner agrarpolitischer Bericht', *Berichte über Landwirtschaft,* Berlin 1939.
Stackelberg, J. von. 'Das Ergebnis der Landwirtschaft', *Wirtschafts Jahrbuch,* 1939.
Stenkhoff, G. 'Landwirtschaftliche Berufs- und Fachschulen im Dienste der Nachwuchssicherung', *Die Deutsche Berufserziehung,* November 1942.
Syrup, Friedrich. 'Arbeitseinsatz gegen Landflucht', *Odal,* Heft 7, July 1939.
Thyssen, Thyge. 'Schleswig-holsteinisches Bauerntum zwischen den beiden Weltkriegen', *Jahrbuch des Angler Heimat Vereins,* 1963-64.
Tönnsen, Max. 'Rückschau und Erinnerungen', *Jahrbuch des Angler Heimat Vereins,* 1963-64.
Treue, Wilhelm. 'Hitlers Denkschrift über den 4 Jahr Plan', *VJH,* 1955.
Walter, A. 'Landwirtschaftliche Zusammenarbeit in Europa', *Berichte über Landwirtschaft,* Vol. 28, 1943.
Woltenmath, K. 'Die historischen Quellen des Erbhofgesetzes und seine Probleme', *Schmollers Jahrbuch für Gesetzgebung, Verwaltung und Volkswirtschaft im Deutschen Reiche,* Leipzig 1939.
Weber, Wilhelm. 'Reichsregierung und Agrarpolitik in der Republik von Weimar 1920-1932', *Berichte über Landwirtschaft,* Vol. 45, 1967 (Reprinted).
Wegener, C. A. 'Die europäische Milch- und Fettswirtschaft', Ibid., Vol. 28, 1939.
Wensdorff, H. 'Das Festpreissystem als Mittel der Erzeugungsschlacht in der Provinz Pommern', Ibid., Vol. 24, 1939.
Woermann, E. 'Europäische Nahrungswirtschaft' etc., *Nova Acta Leopoldina,* Vol. 14, No. 99, Halle 1944.

MISCELLANEOUS SOURCES

Arbeitsmaiden in der Nordmark. Berlin 1943, RAD.
Entscheidungen des Reichserbhofgerichthofs. Berlin, 1935-39.
Foreign Trade and Exchange Controls in Germany. Washington, 1942.

Grundlagen, Aufbau und Wirtschaftsordnung des NS Staates. Spaeth & Linde Verlag, Berlin 1938.

Hajek, Hans. (Ed.). *Landwirtschaftliche Berufe für Jungen.* Reichselternwarte Verlag, Berlin 1941.

Jacoby, Kurt (Ed.). *Das Dritte Reich im Aufbau.* Vol. 1 Quelle & Meyer, Leipzig 1938. Vol. 2 & 3, 1939.

Kaufmann, Günter. *Der Reichsberufswettkampf.* Junker & Dünnhaupt Verlag, Berlin 1935.

Landwirtschaftliche Statistik, 1937 and 1938. RNS Verlag, Berlin.

Rassenpolitik. SS Berlin, probably 1942.

Rassenpolitik im Kriege. Gau South Hanover-Brunswick, Hanover 1941.

Reichsgesetzblatt. Ministry of the Interior, Berlin 1933-45.

Reichsministerialblatt der landwirtschaftlichen Verwaltung. Ministry of Food & Agriculture, Berlin, 1936-43.

Statistisches Jahrbuch des Deutschen Reiches. Reichsamt für Statistik, Berlin 1932, 1936 and 1941-42.

Statistisches Handbuch von Deutschland. Munich 1949.

Vierteljahrsheft zur Statistik des Deutschen Reiches. Berlin 1938.

NOTES, *Chapter 1*

1. M. Kater, 'Zur Soziographie der frühen NSDAP', April 1971, *VJH*, pp. 138ff.
2. The proposals are described in R. Kühnl, 'Zur Programmatik der NS Linken: Das Strasser Program von 1925-6', 1966 *VJH*.
3. See W. Treue, 'Hitlers Denkschrift über den 4 Jahresplan', 1955 *VJH*.
4. See *Trial Brief für den Angeklagten Richard Walther Darré*, handed in at Nuremberg 1948 by Dr H. Merkel, p. 222.
5. K. Verhey, *Der Bauernstand und der Mythos von Blut und Boden, mit besonderer Berücksichtigung auf Niedersachsen* (henceforth Verhey) doctoral dissertation, Univ. of Göttingen 1965, p. 103.
6. J. Noakes, *The Nazi Party in Lower Saxony 1921-1933*, OUP, London 1971, p. 101.
7. Reichskommissariat für die Ueberwachung der öffentlichen Meinung, report LA 309/23055.
8. G. Seifert, 'Beginn und Entwicklung der ersten norddeutschen Kämpfe der NSDAP in Hannover und Niedersachsen', NSA 3101 GI.
9. Noakes, 1971, op. cit., Note 6, pp. 105-6.
10. Hannover Landbund 15 December 1928.
11. Reinhardt to Himmler 5 March 1929, ADC Himmler.
12. These differences are summarised from G. Stoltenberg, *Politische Strömungen im Schleswig-holsteinischen Landvolk 1918-1933*, Droste Verlag Düsseldorf 1962, pp. 31ff.
13. Regpräsident to Prussian Ministry of Interior 30 September 1929, LA 309/22784.
14. 'Bei K-Husum am 10.9.1929 beschlagnahmtes Material'. Ibid for the police report. Unless otherwise stated, information for this account comes from local police reports, LA 309/22696, 22668, 22784 and 22998.
15. Stoltenberg, 1962, op. cit., Note 12, p. 98.
16. Reichskommissariat für die Ueberwachung der öffentlichen Meinung report April 1926, LA 309/22784.
17. H. Beyer, 'Die Landvolkbewegung Schleswig-holsteins und Niedersachsens 1928-32', *Jahrbuch für Kreis Eckernförde*, 1957, p. 174.
18. E. Topf, *Die Grüne Front*, Rohwohlt Verlag, Berlin 1933, pp. 5ff, for an account of these incidents.
19. T. Thyssen, 'Schleswig-holsteinisches Bauerntum zwischen den beiden Weltkriegen', *Jahrbuch des Angler Heimatvereins, 1963-64*, p. 152.
20. See H. Beyer, 'Das Bauerntum Angelns während der grossen Krise 1927-32', *Jahrbuch des Angler Heimatvereins*, 1961-62, p. 176.
21. Beyer, 1957, op. cit., Note 17, p. 186.
22. Stoltenberg, 1962, op. cit., Note 12, p. 133.
23. T. Thyssen, *Bauer und Standesvertretung*, Karl Wachholtz, Neumünster,

1958, p. 406ff.

24. Beyer, 1961-62, op. cit., Note 20, p. 159.

25. Stoltenberg, 1962, op. cit., Note 12, p. 143.

26. D. Orlow, *The History of the Nazi Party 1919-1933,* Univ. of Pittsburgh, 1969, p. 144.

27. Vide police report 21 March 1929 for Heide, LA 309/22696, and Landrat Meldorf to Regpräsident 27 November 1929, LA 309/22998. See also Heberle, *Landbevölkerung und Nationalsozialismus,* Stuttgart 1963, p. 39.

28. M. Kater, 1971, op. cit., Note 1, p. 143.

29. J. A. Beck (Ed.), *Kampf und Sieg,* Gau, South Westphalia 1934, p. 80.

30. For Bavaria Reinhardt to Himmler 5 March 1929, ADC Himmler. For Lower Saxony monthly reports to Gauleiter April 1932 from Nienburg and Alfeld, NSA 310 I B11.

31. Correspondence March-May 1929, LA 309/22784.

32. Stoltenberg, 1962, op. cit., Note 12, p. 164.

33. Heberle, 1963, op. cit., Note 27, p. 116.

34. For a fuller discussion of this theme see H. A. Winkler, 'Extremismus der Mitte? ', 1972 *VJH,* pp. 175ff.

NOTES, *Chapter 2*

1. Agricultural Chamber Schleswig-Holstein to Oberpräsident 22 October 1931, LA 301/4111.

2. E. Topf, 1933, op. cit., ch. 1, Note 18, p. 192.

3. J. B. Holt, *German Agricultural Policy 1918-1934,* Univ. of Carolina Press 1936, p. 163.

4. For details H. Haushofer, *Ideengeschichte der Agrarwirtschaft und Agrarpolitik,* Vol. 2, Bayerischer Landwirtschaftsverlag Munich 1958, pp. 170ff; and K. Verhey, op. cit., Ch. 1, Note 5, p. 31ff.

5. E. Topf, 1933, op. cit., Note 2, p. 142, and ADC Darré.

6. Letter 31 December 1929, ADC Darré.

7. R. W. Darré, *Das Bauerntum als Lebensquell der nordischen Rasse,* Lehmanns Verlag, Munich 1928.

8. R. W. Darré, *Neuadel aus Blut und Boden,* Lehmanns Verlag, Munich 1929.

9. H. Beyer, 1961-62, op. cit., ch. 1, Note 20, p. 161.

10. H. Gies, 'NSDAP und landwirtschaftliche Organisation in der Endphase der Weimarer Republik', *VJH* 1967, pp. 343f and Darré's own correspondence ND 142, 139, 140, 148 and 91 for this account, unless otherwise stated.

11. Verhey, op. cit., ch. 1, Note 5, p. 31.

12. This account is based on Gies, 1967, op. cit., Note 10, unless otherwise stated.

13. Bericht über die Tätigkeit der landwirthschaftlichen Abteilung der

NSDAP im Gau Sachsen, ND 140.

14. Stoltenberg, 1962, op. cit., ch. 1, Note 12, p. 158.

15. H. Kretschmar, *Deutsche Agrarprogramme der Nachkriegszeit,* Junker and Dünnhaupt, Berlin 1933, p. 80.

16. NS Freiheitskämpfer 11 February 1931.

17. Noakes, 1971, op. cit., ch. 1, Note 6, p. 166.

18. Gies, 1967, op. cit., Note 10, p. 359: by July 1932 he was an NSDAP deputy in the Reichstag.

19. Noakes, 1971, op. cit., ch. 1, Note 6, p. 167.

20. Gies, 1967, op. cit., Note 10, pp. 353-4 for an account in more detail.

21. Thyssen, 1958, op. cit., ch. 1, Note 23, pp. 250; 247.

22. Stoltenberg, 1962, op. cit., ch. 1, Note 12, p. 177.

23. Rügensches Kreisblatt, 6 November 1932.

24. M. Broszat, *Der Staat Hitlers,* Deutschertaschenbuch Verlag, Munich 1966, p. 12.

25. ND 140, op. cit., Note 13.

26. Noakes, 1971, op. cit., ch. 1, Note 6, p. 126.

NOTES, *Chapter 3*

1. *Landwirtschaftliche Statistik 1937,* published by Administration Section of the Reichsnährstand, p. 100, and *WB* 29 April 1931.

2. From RM 8,728 million in 1926 to RM 11,425 million in 1932. H. Bente, *Landwirtschaft und Bauerntum,* Junker und Dünnhaupt, Berlin 1937, p. 152.

3. In 1931 as many as 5,798 separate holdings were so disposed of, and in the following year 7,060. *Landwirtschaftliche Statistik,* 1938, p. 21.

4. H. Beyer, 'Die Agrarkrise und das Ende der Weimarer Republik', *Zeitschrift für Agrargeschichte und Agrarsoziologie,* April 1965, p. 80.

5. Letter from Dr Heim 28 September 1927, BA R431/1301.

6. For the debate over the Act see J. B. Holt, 1936, op. cit., ch. 2, Note 3, p. 18ff., and Stoltenberg, 1962, op. cit., ch. 1, Note 12, pp. 70ff.

7. W. Goerlitz, *Die Junker. Adel und Bauer im deutschen Osten,* C. A. Starke Verlag, Limburg 1957, p. 357, and union letter 26 November 1925, LA 301/4089.

8. Quoted in W. Treue (Ed.), *Deutschland in der Weltwirtschaftskrise in Augenzeugen Berichten,* Karl Rauch Verlag, Düsseldorf 1957, p. 265.

9. See Haushofer, 1958, op. cit., ch. 2, pp. 182ff, and Holt, 1936, op. cit., ch. 2, pp. 118-19.

10. *WB* 8 April 1931.

11. Ibid 18 February 1931.

12. For a summary see Haushofer, 1958, op. cit., ch. 2, Note 4, p. 181, and Holt, 1936, op. cit., ch. 2, Note 3, p. 119.

13. cf. von Kalckreuth's remark that the German people must be compelled to eat German food. E. Topf, 1933, op. cit., ch. 1, Note 18, pp. 184-85.

14. *Goslarsche Zeitung* 15 January 1933, *WB* 23 January 1933, and Letter 12 January 1933, BA R43II/196.

15. E. Borsig, *Reagrarisierung Deutschlands?* Gustav Fischer Verlag, Jena 1934, p. 161.

16. Statistically he (Treviranus) was correct. In May of that year the price in Chicago was RM 128 per ton, in Berlin RM 285/287. *WB* 20 May 1931.

17. For example, only 21 percent on butter and approximately 50 percent on pigs and pork, but 180/200 percent on grain and 326 percent on wheat flour.

18. Vide G. Kroll, 'Die deutsche Wirtschaftspolitik in der Weltwirtschaftskrise', in K. E. Born (Ed.), *Moderne deutsche Wirtschaftsgeschichte,* Kippenheuer and Witsch, Cologne 1966, pp. 400-1.

19. Beyer, 1965, op. cit., Note 4, p. 85.

20. *WB* 6 May and 10 June 1931 and Agricultural Chamber Schleswig-Holstein letters 13 January 1932, LA 301/4111.

21. Prussian Ministry of Interior to Oberpräsident 17 November 1931, LA 301/4111.

22. Agricultural Chamber report 26 April 1932, LA 301/4036.

23. Regional Ministries of Justice and Economics letters 4 December 1931, HSAS 980/E130 IV and 30 November 1931, HSAS 452/E130 IV.

24. For Bavarian reactions Berliner Tageblatt article BA NS26/962: for Württemberg Ministry of State letter 28 December 1932, HSAS 913/E130 IV.

25. W. F. Boyens, *Die Geschichte der ländlichen Siedlung,* Vol. II, Landschriften Verlag, Berlin/Bonn 1959, pp. 192-93.

26. O. Braun, *Von Weimar zu Hitler,* Harmonia Norddeutsche Verlagsanstalt, Hamburg 1949, p. 390.

27. D. Petzina, 'Hauptprobleme der deutschen Wirtschaftspolitik 1932-3', *VJH* 1967, for an account of von Papen's policy.

28. *WB* 7 December 1932.

29. Cabinet minutes and correspondence January 1933, BA R43II/192.

30. The formal law authorising this came out in December: Holt, 1936, op. cit., ch. 2, Note 3, pp. 120-121.

31. NS Kurier 29 December 1932 and Petzina, 1967, op. cit., Note 27, pp. 36-37.

32. *VB* 3 January 1933 and D. Hertz-Eichenrode, *Politik und Landwirtschaft in Ost Preussen 1918-1930* West Deutscher Verlag, Cologne 1969, p. 75.

33. Gies, 1967, op. cit., ch. 2, Note 10, p. 374.

34. The cabinet was currently being bombarded by missives from commerce and industry protesting about a possible debt-moratorium for agriculture, BA R43II/192.

35. *VB* 17 January 1933, and ND 159.

36. Official report. 21 April 1933, GSA 1112.

37. Secretary Baden Centre Party to Wirth October 1922, BA R43I/1277.

38. W. Domarus, *Hitler: Reden und Proklamationen,* Vol. II, Süddeutscher Verlag, Munich 1965, pp. 93-94. Henceforth 'Domarus'.

39. *VB* 17 January 1932.

40. *West Deutscher Beobachter,* 23 March 1930.

41. *VB* 5 June 1931, and Speech HSA(S) 535/E.130/IV.

42. Noakes, 1971, op. cit., ch. 1., Note 6, p. 168.

43. K. D. Bracher, *Die Auflösung der Weimarer Republik,* Schwarzwald Ring Verlag, Villingen 1964, p. 115.

44. Letters to party HQ and LGFs October-November 1932. ADC Darré and ND 142.

45. To LGFs 20 October 1937, ibid.

46. Beyer, 1961-62, op. cit., ch. 1, Note 20, p. 160, note 28.

47. *VB* 22 January 1932.

48. Circular to LFGs 11 May 1932, ND 145, and monthly activities report Alfeld (Lower Saxony) April 1932, NSA 310I B11.

49. *VB* 9 February 1932.

50. Details in ND 145.

51. Circular 6 November 1932, ibid for these instructions.

52. Monthly report Nienburg (Weser), NSA 310I B11.

53. *WB* 1 January 1930.

54. Richtlinien für die Arbeit unter den Bauern, rural secretariat of the KPD's Central Committee 20 February 1925, LA 309/23055.

55. For the figures Gies, 1967, op. cit., ch. 2, Note 10, p. 355.

56. Vide *VB* 10 July 1930 for exact figures.

57. Quoted in A. Nicholls, 'Hitler and the Bavarian background to National Socialism' in A. Nicholls and E. Matthias (Eds.), *German Democracy and the triumph of Hitler,* Allen & Unwin, London 1971, p. 111.

58. Landrat Schleswig report 5 August 1930, LA 309/22998.

59. Report for Hameln-Bad Pyrmont, June 1932, NSA 3101 B11.

60. Similarly the NSDAP possessed far fewer newspapers than their opponents, which again emphasises the significance of purely local leadership.

NOTES, *Chapter 4*

1. BA R43II/196 and BA R43II/203 February 1933.

2. *WB* 19 April 1933.

3. F. Jacobs, *Von Schorlemer zur Grünen Front,* Hans Altenberg Verlag, Düsseldorf 1957, p. 53 and pp. 77-78.

4. Circular to LGFs 20 March 1933, ND 140, and BA R43II/203.

5. Jacobs, 1957, op. cit., Note 3, pp. 79-87.

6. Circular to LGFs 11 April 1933, ND 140.

7. Stoltenberg, 1962, op. cit., ch. 1, Note 12, pp. 190-2 and Jacobs, 1957, op. cit., Note 3, p. 85.

8. Minutes 4 April 1933, BA R43I/1461.

9. HSAS 653 E130/IV.

10. Thyssen, 1958, op. cit., ch. 1, Note 23, p. 252 and M. Tönnsen, 'Rückschau und Erinnerungen', Jahrbuch des Angler-Heimat Vereins, 1963-64, p. 138.

11. Eroberung des landwirtschaftlichen Genossenschaftwesens, 28 February 1933, ND 140.
12. 13 April 1933, ibid.
13. M. Broszat, *Der Staat Hitlers,* Deutscher Taschenbuchverlag, 1966, p. 231.
14. According to Dr Reischle in an interview with the author.
15. *WB* 7 June 1933.
16. Domarus, 1965, op. cit., ch. 3, Note 38, p. 253.
17. Broszat, 1966, op. cit., Note 13, p. 231.
18. 29 March 1933, ND 140.
19. Letter to LGFs 11 April 1933, ND 141.
20. Quoted in M. Goehring, *Alles oder Nichts* JCB Mohr (Paul Siebeck), Tübingen 1966, p. 70.
21. *WB* 31 October 1933.
22. Cabinet minutes 16 February 1933, and 2 March 1933, BA R43I/1459 and Hugenberg to Cabinet 18 March 1933, BA R43I/1460.
23. Verordnung . . . zur Förderung der Landwirtschaft, 25 February 1933, *RGB*(I) p. 80 and WTB.
24. Cabinet minutes for March 1933, BA R43I/1460.
25. Von Rohr to Lammers 16 June 1933, BA R43II/199.
26. REM reports 20 February 1933 and 20 July 1933, BA R2/18202.
27. Chancellery to Hugenberg 31 May 1933, BA R43II/192.
28. Bente, 1937, op. cit., ch. 3, Note 2, p. 177.
29. Gesetz zur Regelung der landwirtschaftlichen Schuldverhältnisse, *RGB*(I), p. 331.
30. Letters 29 March and 1 April 1933, ND 140.
31. *WB* 21 June 1933.
32. Minutes 31 May 1933, BA R43I/1462.
33. H. Haushofer, *Die deutsche Landwirtschaft im technischen Zeitalter,* Verlag Eugen Ulmer, Stuttgart 1963, p. 261.
34. Cabinet minutes for February/March 1933, BA R43I/1459.
35. BA R43II/192, and 20 March 1933, ND 140.
36. Minutes 28 April 1933, BA R43I/1461.
37. Vide A. Ritthalter, 'Eine Etappe auf Hitlers Weg zur ungeteilten Macht', *VJH* 1960, p. 199: the account here is based on this article unless otherwise stated.
38. K. D. Bracher, et al., *Die Nationalsozialistische Machtergreifung,* West Deutscher Verlag, Cologne 1960, pp. 188 and 575.
39. Minutes BA R43I/1463.
40. Ritthalter, 1960, op. cit., Note 37, pp. 199-200.
41. Gies, 1967, op. cit., ch. 2, Note 10, p. 375.
42. It has been suggested that Hindenburg had desired the DNVP's leader in the cabinet. Bracher 1964, op. cit., ch. 3, Note 43, p. 727.
43. IfZ ZS 1622: unfortunately Willikens gave no date.
44. Ritthalter, 1960, op. cit., Note 37, gives an account of their interview, pp. 208-14.
45. E. Matthias and R. Morsey, *Das Ende der Parteien 1933,* Droste Verlag, Düsseldorf 1960, pp. 613-14.

NOTES, *Chapter 5*

1. Correspondence, etc. ND 128 and Dr Reischle in an interview with the author.
2. H. Reischle, *Die deutsche Ernährungswirtschaft,* Junker and Dünnhaupt Verlag, Berlin 1935, p. 5.
3. H. Backe, 'Grundsätze einer lebensgesetzlichen Agrarpolitik', in H. Reischle (Ed.), *Deutsche Agrarpolitik,* Berlin 1934, p. 71.
4. Ibid p. 65 and R. W. Darré, 'Bauer und Landwirt', ibid., p. 46.
5. R. W. Darré, 'Das Ziel', ibid., p. 16.
6. Regelung des ständischen Aufbaues, *RGB*(I), p. 495.
7. Vide Darré to Chancellery 8 September 1933, BA R43II/201.
8. Vorläufigen Aufbau, etc. *RGB*(I), p. 627.
9. Minutes 11 July 1933, BA R43I/1465.
10. The classic source for the RNS is H. Reischle and W. Saure, *Der Reichsnährstand-Aufbau, Aufgaben und Bedeutung,* Berlin 1936.
11. *WB* 26 July 1933.
12. Von Rohr's proposals in draft April 1933, ND 143.
13. *Vierteljahrsheft zur Statistik des deutschen Reiches,* Pt II, 1939, p. 36.
14. *Schwäbische Zeitung,* 22 July 1925.
15. Landwirtschaftliche Hochschule Stuttgart report HSAS 646 E/130 IV.
16. Haushofer, 1958, op. cit., ch. 2, Note 4, p. 104. The *Landbund* and the Agrarian Council had repeatedly called for similar reforms. Holt, 1936, op. cit., ch. 2, Note 3, p. 209.
17. Correspondence in April 1925, HSAS 343 E/130 IV.
18. Draft Bill and correspondence February 1928, HSAS 343 E/130 IV.
19. For Hanover K. Woltenmath, 'Die historischen Quellen des Erbhofgesetzes und seine Probleme', *Schmollers Jahrbuch für Gesetzgebung, Verwaltung und Volkswirtschaft im Deutschen Reiche,* 1939.
20. Minutes 26 September 1933, BA R43I/1465.
21. Quoted in D. Schoenbaum, *Hitler's Social Revolution,* Weidenfeld & Nicolson, London 1966, p. 45.
22. Bente, 1937, op. cit., ch. 3, Note 2, p. 172.
23. Ibid for all figures here pp. 183 and 187. Taxes to agriculture were cut by RM 250 million yearly in autumn 1933.
24. W. Bauer and P. Dehen, 'Landwirtschaft und Volkseinkommen', *Vierteljahrsheft zur Wirtschaftsforschung,* Heft 3, 1938-39, p. 427.
25. *WB* 13 September 1933.
26. Domarus, 1965, op. cit., ch. 3, Note 38, p. 304.
27. Speech to Raiffeisen Society June 1938.
28. Winkler, 1972, op. cit., ch. 1, Note 34, p. 190.
29. In 1937-38 the board for dairy products handled 11.5 percent of butter consumed in Germany. Reports February 1936 and March 1939, BA R2/18075.

NOTES, *Chapter 6*

1. Dr Reischle head of the Staff Office in an interview with the author.
2. Lbsch. Westphalia to Darré 3 May 1937, ADC Reichsnährstand Habbes/Eltz-Rübenach.
3. Internal circular 1 February 1935 HSAD.
4. Vide Beyer, 1961-62, op. cit., ch. 1, Note 20, p. 152, and Thyssen, 1958, op. cit., ch. 1, Note 23, pp. 297-98.
5. Regpräsident Osnabrück to Prussian Ministry of Agriculture and Regpräsident Münster July 1934, BA R43II/193.
6. KBF Hameln-Bad Pyrmont to Lbsch. Lower Saxony 5 February 1940, NSA 331 B61.
7. According to Professor Haushofer in an interview with the author.
8. Thyssen, 1958, op. cit., ch. 1, Note 23, p. 299.
9. See letter 22 July 1935, LUD 1645/K710.
10. Kbsch. Niederberg (Rhineland) to OBF Mettmann 29 September 1936, HSA D130.
11. Letters 21 November 1936 and 25 March 1938, ibid.
12. KBF Bad Segeberg to Lbsch. 20 July 1939, LA 691/430.
13. Letter 13 May 1937, NSA 331 B1.
14. OBF Ruhpolding (Bavaria) to KBF Traunstein 19 January 1938, BA R2/18291.
15. Tätigkeitsbericht Kbsch. Niederberg 20 June 1936, HSAD 127.
16. Letters 28 February 1936 and 3 November 1937, NSA 331 B35 and Unna Hemmerde.
17. Darré to Goering 2 June 1938, BA R43II/202a.
18. BA R43II/193.
19. Letter to REM 29 June 1935, BA R43II/203.
20. Industry and Trade Chamber Königsberg to Ministry of Economic Affairs, BA R43II/203, undated minute (1934?) HSAS 653/E 130 IV, and Schacht to Darré 28 March 1935, BA R43II/332.
21. *Die Landesbauernschaften in Zahlen,* RNS Folge 2, Berlin 1938, p. 122.
22. Verordnung der Ordnung der Getreidewirtschaft *RGB*(I), pp. 629ff.
23. C. W. Guillebaud, *The Economic Recovery of Germany 1933-8,* Macmillan, London 1939, p. 162.
24. Report to REM 2 March 1937, BA R2/18155.
25. *How National Socialist Germany is solving her agricultural problems,* Berlin 1936, rNS, p. 21.
26. H. Reischle, 'Grundlagen, Aufbau und Wirtschaftsordnung des NS Staates', in G. Feder (Ed.), *Der deutsche Staat auf nationaler und sozialer Grundlage,* Vol. II, Verlag Franz Eher, Munich 1938, p. 9.
27. W. Meinhold, *Grundlagen der landwirtschaftlichen Marktordnung,* Paul Parey, Berlin 1937, p. 87.
28. *Die Landesbauernschaften in Zahlen,* p. 145.
29. Correspondence January-February 1935, NSA 331 B21.

30. B. Vollmer, *Volksopposition im Polizeistaat,* Deutsche Verlags Anstalt, Stuttgart 1957, pp. 44-45 and 60-62, and *Frankfurter Zeitung* 21 May 1937.
31. Correspondence July 1936 and May 1937, HSAD 1067, and 19 January 1938, BA R2/18291.
32. Minutes 26 November 1934, NSA 122a XXXII 80.
33. 1 November 1936, ND 146.
34. Vollmer, 1957, op. cit., Note 30, pp. 60-62 and BA R43II/193.
35. Letters January 1938, BA R2/18291.
36. Report February-March 1938, HSAD 1069.
37. Letters 30 September and 9 October 1934, BA R43II/193.
38. Vollmer, 1957, op. cit., Note 30, pp. 60-62..
39. Thyssen, 1958, op. cit., ch. 1, Note 23, p. 300, and report 20 November 1934, NSA 122a XXXII, 80.
40. Vollmer, 1957, op. cit., Note 30, p. 62.
41. *Foreign Trade and Exchange Controls in Germany* (henceforth FTEC), US Tariff Commission Washington 1942, p. 61.
42. Correspondence and minute July 1934, BA R43II/193.
43. G. Bry, *Wages in Germany,* OUP, London 1960, p. 425.
44. Report BA R43II/318.
45. Minutes BA R43I/1470.
46. *Berliner Tageblatt* 7 November 1934.
47. Goerdeler to Hitler 21 December 1934 ibid.
48. *Berliner Tageblatt* 9 November 1934, *Frankfurter Allgemeine Zeitung* 16 November 1934, and A. E. Simpson, *Hjalmar Schacht in perspective,* Mouton, The Hague 1969 pp. 90-91.
49. Letter 1 November 1936, ND 146.
50. Letter 5 September 1936, ADC Darré.
51. Vide Lammers to Mayor of Stuttgart 31 July 1935, BA R43II/317.
52. Diener des Volkes, *VB* 1 January 1935.

NOTES, *Chapter 7*

1. Reports July and November 1934, NSA 122a XXXII 80.
2. Minutes 3 April 1935, ibid.
3. Regpräsident Cologne report October-November 1937, HSAD 1068.
4. Ministry of Interior to regional governments 8 January 1935, HSAS 913 E/130 IV.
5. Letter 31 October 1934, HSAS 653 E/130 IV.
6. Minutes 14 June 1935, ADC Darré.
7. Letter 6 October 1934, BA R2/18291.
8. Correspondence and conference minutes, HSAS 653 E/130 IV.
9. Regional government minutes, 17 July 1937, ibid.
10. Letter to deputy LBF Bavaria 3 March 1937, ADC Reichsnährstand.

11. Correspondence January-February 1934, HSAS 653 E/130 IV.
12. Gesetz über die landwirtschaftlichen Schulen draft 7 May 1938, BA R2/18187 and correspondence August-September 1942, ibid.
13. Ministry of Education Württemberg to Ministry of State 22 February 1934, HSAS 653 E/130 IV.
14. Correspondence 19-20 November 1934, NSA 122a XXXII 80.
15. Correspondence December 1930 – January 1931, ADC Darré.
16. Herr von Rohr in an interview with the author.
17. ND 142, Gies, 1967, op. cit., ch. 2, Note 10, p. 349 and letter 13 June 1934, BA R43II/203.
18. Correspondence June-July 1934, BA R43II/203 and ADC Karpenstein.
19. M. Broszat, 'The Concentration Camps 1933-45', in H. Krausnick (Ed.), *The Anatomy of the NS State,* Collins, London 1968, p. 408.
20. September 1933 report ADC Angelegenheit Ost Preussen, *VB* 15 March 1933, and letter 27 July 1933, IfZ Fa508.
21. Unless otherwise stated this account is based on material in IfZ 508.
22. Correspondence July-October 1933, ADC D932/3.
23. Meinberg official report 28 September 1933, ADC Angelegenheit Ost Preussen: this is now the source unless otherwise stated.
24. Letter 16 April 1934, BA R43II/207.
25. See Darré to Hitler 13 June 1934, BA R43II/203.
26. Correspondence July-September 1938, BA R43II/194.
27. *Trial Brief,* p. 217.
28. Topf, 1933, op. cit., ch. 1, Note 18, p. 168.
29. Vide speech by Dr Steiger, Minister of Agriculture in Prussia, February 1930, LA 301/4089.
30. *Die landwirtschaftlichen Verhältnisse in der Landesbauernschaft Weser-Ems,* 1937, published by Lbsch., p. 43; copy in NSA 122a XXXII 88a.
31. Wochenblatt der Landesbauernschaft Württemberg, 25 May 1935.
32. Arbeitertum, 1 July 1937.
33. Undated list for Kbsch. Niederberg (Rhineland) HSAD 127.
34. 19 October 1935, NSA 331 B35, and letter 3 March 1936, HSAD 127.
35. KBF Niederberg to Lbsch. 28 May 1936, ibid.
36. F. Wunderlich, *Farm Labour in Germany 1810-1945,* Princeton Univ. Press 1961, p. 217: in 1939 the proportion for agriculture was higher still, 32 cases out of 120, ibid.
37. *VB* 11 January and 28 February 1935 and report May 1937, BA R43II/528.
38. Darré's circular to Ministries, 27 September 1935, BA R2/18291.
39. Correspondence June 1936 HSAD 127.
40. Correspondence February-April 1937, BA R43II/528.
41. Trustee of Labour reports May and July 1937: a similar situation was reported for Mecklenburg, ibid.
42. Correspondence February-March 1938, BA R43II/529.
43. Correspondence May-September 1938, BA R43II/194.
44. L. Weiss, *Die Abmeierung,* Hans Buske Verlag, Leipzig 1936, p. 1.
45. M. Frauendorffer, 'Deutsche Arbeitsfront und ständischer Aufbau', *NS*

Monatsheft, No. 54, September 1934, p. 819.

46. Vide *Trial Brief,* p. 217.

47. Quoted in Schoenbaum, 1966, op. cit., ch. 5, Note 21, p. 129.

48. Vide Goering to Darré 27 October 1936, ND 146 for this attitude.

49. Unless otherwise stated this information is from correspondence and reports in March-June 1937, ADC Meinberg.

50. Eltz-Rübenach to Darré 26 April 1937, ADC Eltz-Rübenach/Habbes.

51. Correspondence April 1937-January 1938, ADC SS Habbes/ Eltz-Rübenach.

52. Darré to Ministry of Finance 25 November 1939, BA R43II/202a.

NOTES, *Chapter 8*

1. ADC Reichsnährstand Backe.

2. *WB* 5 December 1928.

3. Thyssen, 1958, op. cit., ch. 1, Note 23, p. 278.

4. For Wagemann's part see Kerrl's oration at his funeral 11 December 1933, GSA p. 133/473.

5. For the old Hanoverian law Woltenmath, 1939, op. cit., ch. 5, Note 19.

6. *WB* 26 July 1933, and minutes 20 September 1933, HSAS 653 E/130 IV.

7. Reichserbhof Court 12 February 1937, BA R22/2248.

8. For Westphalia reaction *WB* 17 and 24 May 1933; for Württemberg minutes 20 September 1933, HSAS 653 E/130 IV.

9. E. Molitor, *Deutsches Bauern- und Agrarrecht,* Quelle und Meyer Verlag, Leipzig 1936, p. 51.

10. Lbsch. Rhineland to RNS Admin. Section 6 May 1936, HSAD 1043.

11. Vierte Durchführungsordnung, Article V, 21 December 1936.

12. Vide R. W. Darré, 'Geburtminderung durch das Erbhofgesetz? ', *Die Dorfkirche,* Heft 5, May 1934, p. 147.

13. HSAD 51 and 10 October 1937, LUD Bundle 210/F. 277.

14. K. Blomeyer, 'Neuerungen im Erbhofrecht', *Jahrbücher für National-ökonomie und Statistik,* 1937, p. 451.

15. *Vierteljahrsheft zur Statistik des Deutschen Reiches,* 1939, Sec. II, pp. 37-38.

16. *WB* 26 July 1933 for summary of an article by Darré on this issue.

17. *Vierteljahrsheft zur Statistik des Deutschen Reiches, 1939,* Sec. II, p. 36.

18. Letter to Chancellery 29 September 1933, BA R43I/1301.

19. Von Rohr, 'Beitrag zur deutschen Agrarpolitik', BA Kl. Erw. 404, pp. 32ff.

20. H. Haushofer and H. J. Recke, *50 Jahre: Reichsernährungsministerium-Bundesernährungsministerium,* Bundesministerium für Ernährung, Landwirtschaft und Forsten, Bonn 1966, (Henceforth, *50 Jahre*).

21. LUD Bundle 211/F 277, and 18 September 1935, BA R22/2248.

22. *RGB*(I), 1933 pp. 749 and 1096, and *RGB*(I), 1936 p. 1069.

23. Dr Hopp, *Deutsche Justiz,* No. 1324, BA R22/2183, 1935.

24. Court Künzelsau to KBF Hohenlohen 6 November 1935, LUD Bundle 210/F277 and *Deutsche Justiz,* No. 1324, 1935.

25. Dr Hopp, 1935, op. cit., Note 23.

26. *WB* 24 May 1933 and minutes 20 September 1933, HSAS 653 E/130 IV.

27. 24 January 1929, LA 309/22696.

28. 30 October 1937, BA R22/2129. See also Regierungsrat Hildesheim letter 2 March 1936, NSA 122a-13 and 24 January 1938, ADC Reichsnährstand Wagner.

29. Beyer, 1965, op. cit., ch. 3, Note 4, p. 64, and report BA R43II/193.

30. Vide Meinberg, 'Diener des Volkes', *VB* 1 January 1935.

31. A. M. Koeppens, *Das deutsche Landfrauenbuch,* Reichsnährstand Verlag, Berlin 1937, pp. 92-93.

32. Ministry of Justice to REM 2 December 1938, BA R22/2129.

33. Correspondence 7 May and 1 September 1938, HSAD 1012.

34. Dr Petersen, *Deutsche Zeitung,* 17 June 1934.

35. Vollmer, 1957, op. cit., ch. 6, Note 30, p. 63.

36. Correspondence November 1933, BA R43I/1301.

37. Beyer, 1965, op. cit., ch. 3, Note 4, p. 64, and letters November-December 1934, BA R43I/1301.

38. According to Professor Haushofer in an interview with the author.

39. See *WB* 31 May 1933 and minute 27 April 1933, BA R43II/193.

40. Woltenmath, 1939, op. cit., ch. 5, Note 19, pp. 35-38.

41. Ibid., p. 49.

42. For the 5-20 Ha category the total number rose between 1882 and 1933 by 225,154. Goerlitz, 1957, op. cit., ch. 3, Note 7, p. 343.

43. RM 11.4 billion as against RM 10.1 billion. *Landwirtschaftliche Statistik 1938,* p. 101 and Bente, 1937, op. cit., ch. 3, Note 2, p. 152.

44. Weiss, 1936, op. cit., ch. 7, Note 44, p. 72.

45. 49 Vohwinkel HSAD and Unna Hemmerde.

46. Thyssen, 1958, op. cit., ch. 1, Note 23, p. 278 and K. Heiden, *Geburt des Dritten Reichs,* Europa Verlag, (Zurich 1934), p. 182.

NOTES, *Chapter 9*

1. This account is based on 'Erbhofgesetz' Articles 40-47, *RGB*(I) p. 685ff and a later directive, ibid p. 749ff.

2. W. Harmening, 'Bauerngerichte', BA R22/2248.

3. Molitor, 1936, op. cit., ch. 8, Note 9, p. 112.

4. According to Herr Kahlke, a former judge at Celle, in a personal interview with the author.

5. Correspondence and programme for East Hanover, GSA, p. 133-426.
6. Correspondence August-September 1939, LA 691/430.
7. Gau E. Hanover to Celle 5 February 1938, GSA, p. 133-426.
8. Quoted in Broszat, 1968, op. cit., ch. 7, Note 19, pp. 422-23.
9. Zweite Durchführungsordnung 19 December 1933, *RGB*(I) p. 1096.
10. Niedersächsische Tageszeitung 20 December 1936.
11. BA R22/2248. For similar delays in Westphalia, lawyer's letter 10 October 1937, ibid.
12. GSA p. 133-473.
13. Tagebuch 277 Minute 20 December 1934, BA R22/2248.
14. Correspondence August 1936-July 1938, BA R16/1271.
15. Bavarian Ministry of Justice to Reich Ministry 7 October 1934, BA R22/2248 and Minute IIa 8 298/38, ibid.
16. NSA 331 B12 and decision November 1934, Unna Hemmerde.
17. Reichsamt für Agrarpolitik to Minister President Württemberg January 1938, HSAS 646 E/130 IV.
18. *Entscheidungen des Reichserbhofgerichtes 1936-37,* p. 382 (henceforth *Entscheidungen*).
19. LUD Bundle 211/F277.
20. Anerbengericht Bargteheide 20 May 1935 and 10 October 1939, LA 355/83, 15 October 1938, NSA 331 B12 and 29 October 1936, Unna Norddinker.
21. LUD Bundle 211/F277, 2 December 1938, Unna Afferde, and Lbsch. Bavaria to REM 15 August 1936, BA R22/2131.
22. Correspondence November 1936-April 1937, Unna Hemmerde.
23. *Entscheidungen, 1935-36,* p. 683 and 1936-37, p. 44.
24. Ibid., pp. 264, 292 and 302 and letters December 1936-July 1937, LA 691/430.
25. Anerbengericht Unna 23 June 1937, Unna Hemmerde.
26. Correspondence June 1934-January 1935, Unna Afferde.
27. Correspondence July-November 1936, GSA p. 133-457.
28. *Wiesbadener Tagesblatt* 23 November 1934, GSA p. 133-456.
29. See *NS Landpost* 1 March 1935.
30. Erbhofrechtsverordnung 21 December 1936, Article X.
31. *Entscheidungen, 1937-38, p. 203, Anerbengericht Künzelsau LUD Bundle 210/F277 and Entscheidungen 1936-37,* p. 306.
32. Anerbengericht Gettorf 14 October 1938, LA 691/413, Weiss, 1936, op. cit., ch. 7, Note 44, p. 30. *Entscheidungen, 1936-37,* p. 219, and correspondence April 1936-March 1937, LA 691/430.
33. 9 August 1935, NSA 331 B35, Anerbengericht Bargteheide 26 January 1937, LA 691/406.
34. *West Deutscher Beobachter* 2 October 1937, and BA R2/18019.
35. Anerbengericht Unna 17 October 1935, Unna Hemmerde. Künzelsau 6 March 1934, LUD Bundle 211/F277 and 20 June 1935, LA 355/379.
36. Wunderlich, 1961, op. cit., ch. 7, Note 36, p. 172.
37. LUD Bundle 211/F277 and BA R22/2248.
38. Vide Minute IIa 8 298/38 and Letter 10 October 1937, BA R22/2248.

39. 1 February 1935, BA R43I/1301.
40. Letter 14 January 1935, LUD Bundle 211/F277.
41. *Neuadel aus Blut und Boden,* p. 76.
42. C. von Dietze 'Deutsche Agrarpolitik seit Bismarck', *Zeitschrift für Agrargeschichte und Agrarsoziologie, 1964,* p. 209.
43. Wunderlich, 1961, op. cit., ch. 7, Note 36, p. 172.

NOTES, *Chapter 10*

1. F. C. von Hellermann, *Landmaschinen gegen Landflucht,* Junker und Dünnhaupt Verlag, Berlin 1939, p. 28 and Wunderlich, 1961, op. cit., ch. 7, Note 36, p. 24.
2. Braun, 1949, op. cit., ch. 3, Note 26, pp. 22-23, Topf, 1933, op. cit., ch. 1, Note 18, pp. 269-70 and Wunderlich, 1961, op. cit., ch. 7, Note 36, p. 27.
3. Quoted in J. P. Mayer, *Max Weber and German Politics,* Faber & Faber, London 1956, p. 40.
4. Minuted 22 February 1926 cited in Hertz-Eichenrode, 1969, op. cit., ch. 3, Note 32, p. 313.
5. A. Gerschenkron, *Bread and Democracy in Germany,* Howard Fertig, New York 1966, p. 132, Wunderlich p. 37 and Hertz-Eichenrode p. 328, and Boyens, 1959, op. cit., ch. 3, Note 25, Vol. I, pp. 378-423.
6. C. Horkenbach, *Das Deutsche Reich 1918-1933,* Presse und Wirtschaftsverlag, Berlin 1935, p. 748.
7. Arbeitslosigkeit und Siedlung, August 1932, BA R43I/1465.
8. W. Haas, *Bedeutung, Aufgaben und Durchführung der Neubildung deutschen Bauerntums östlich der Elbe im NS Staat,* Dissertation Heidelberg University 1936, pp. 51-52. Cf. Hitler's speech to the generals 3 February 1933, where he envisages a large-scale programme as the only answer to unemployment.
9. Minutes 26 April 1932, GSA 1112, Oberpräsident to Prussian Ministry of State 26 June 1932, GSA 1111, and W. Treue, 1955, op. cit., ch. 1, Note 3, pp. 104-5.
10. W. Frick (Ed.), 'Die Nationalsozialisten im Reichstag 1924-31', Heft 7, NS Bibliothek 1937, p. 108.
11. Correspondence BA N26/952, and Richtlinien für die Neubildung deutschen Bauerntums, BA R22/2128.
12. Bente, 1937, op. cit., ch. 3, Note 2, p. 74.
13. Neubildung deutschen Bauerntums, *RGB*(I), p. 517.
14. K. Kummer, 'Der Weg deutscher Bauernsiedlung', *Odal,* November 1939, p. 938.
15. Jurda to Darré 15 February 1932, BA NS 26/952, and correspondence June to August 1934, BA NS 26/946.
16. Reichsstelle für Siedlungsplanung to Ministry of State Württemberg 22 December 1933, HSAS 452 E/130 IV.

17. Report 6 April 1933 GSA 1112 and Borsig, 1934, op. cit., ch. 3, Note 15, p. 19.
18. See 'Ostelbier', *Odal* June 1934 which is a reprint of his speech.
19. The *Observer*, 13 May 1934.
20. Correspondence September 1933, BA R43I/1301.
21. Correspondence May-July 1933, IfZ Fa 508.
22. Schoenbaum, 1966, op. cit., ch. 5, Note 21, p. 179.
23. 13 January 1934, BA R43II/193, Minutes 27 April 1933, R43II/192, and Heiden, 1934, op. cit., ch. 8, Note 46, p. 185.
24. Goerlitz, 1957, op. cit., ch. 3, Note 7, p. 404. In 1932-33, 36 percent of all newly-commissioned lieutenants were aristocrats, Schoenbaum, 1966, op. cit., ch. 5, Note 21, p. 254.
25. Kummer to Himmler 27 November 1937, BA NS26/944.
26. Quoted in B. Buchta, *Die Junkers und die Weimarer Republik,* VEB Verlag, Berlin 1959, p. 137.
27. Bracher, 1964, op. cit., ch. 3, Note 43, p. 514.
28. *Vierteljahrsheft zur Statistik des Deutschen Reiches* 1938, Sec. II, pp. 22-23 and undated minute, BA NS26/946.
29. Reichsamt für Agrarpolitik report January 1938, HSAS 656 E130/IV.
30. Verhey, op. cit., p. 77 and letter to Ministry of Finance 11 January 1937, BA R22/2130.
31. From RM 709 to RM 905 per Ha. *Vierteljahrsheft zur Statistik des Deutschen Reiches,* 1938, Sec. III, pp. 22-23.
32. H. Reischle, *Soziale Praxis,* Heft 4, 1938, p. 203ff.
33. Vide Reichsamt report January 1938, HSAS 656 E130/IV.
34. *RMB* 28 May 1936, 26 July and 10 May 1938, *Vierteljahrsheft,* op. cit., Note 28, pp. 4-7, Reichsamt Report, op. cit., Note 33, and E. Molitor, 1936, op. cit., ch. 8, Note 9. p. 173.
35. 19 November 1934, BA NS26/955.
36. A. Rosenberg, *Die Artamanenbewegung,* Undated dissertation Institut für Agrarwesen und Agrarpolitik, Berlin, p. 44, and Lbsch. E. Prussia to REM 20 November 1934, BA NS26/955.
37. Correspondence 31 December 1937, ADC Reichsnährstand Willikens, and 26 February 1938, ADC Reichsnährstand Eltz-Rübenach.
38. *Die Landesbauernschaften in Zahlen,* p. 187.
39. H. J. Seraphim, 'Neuschaffung deutschen Bauerntums', *Zeitschrift für die gesamte Staatswissenschaft,* 1935, p. 153 and *RGB*(I), p. 195.
40. 15 March 1935.
41. *Statistisches Jahrbuch,* 1941-42, p. 386.
42. *Die Landesbauernschaften in Zahlen,* p. 186.
43. Correspondence November 1934, BA NS26/955.
44. Reichsamt für Agrarpolitik report, 1938, op. cit., Note 33.
45. Correspondence July-August 1937, GSA 1112, and Goerlitz, 1957, op. cit., ch. 3, Note 7, p. 391.
46. Undated minute BA NS26/946.
47. Minutes ADC RFSS 1030.
48. H. Höhne, *The Order of the Deaths Head,* Secker & Warburg, London

1969, p. 447, R. L. Koehl, *RKFDV: German resettlement and population policy 1939-1945,* Harvard Univ. Press 1957, pp. 44 and 52.

49. J. Noakes, 'Conflict and development in the NSDAP 1924-7', *Journal of Contemporary History,* Vol. I, No. 4, Sage Publications, 1966, pp. 24-25.

50. ADC Backe for details of his life and career.

51. Vide Dr Vager, 'Gegenwartsprobleme der ländlichen Siedlung', *Deutsche Siedlung,* 5 March 1937 and Schoenbaum, 1966, op. cit., ch. 5, Note 21, p. 168.

52. *Vierteljahrsheft,* op. cit., Note 28, Sec. III, pp. 22-23 and ibid., pp. 4-7

53. *Die Landesbauernschaften in Zahlen,* op. cit., Note 42, p. 185.

54. *Westfälische Landeszeitung* 25 July 1932.

55. Wunderlich, 1961, op. cit., ch. 7, Note 36, p. 179.

56. RNS Admin. Office to Lbsch. Lower Saxony 25 July 1935, NSA 331 B21.

57. Kummer to SS Stettin 5 November 1936, BA NS26/944.

58. *6 Jahre Neubauernauslese* published by RNS 1940, p. 6 (henceforth *6 Jahre*).

59. Kummer to Willikens 4 September 1934, BA NS26/946 and *6 Jahre,* pp. 20-21.

60. Correspondence January-June 1939, BA R22/2129.

61. J. von Schultze, *Deutsche Siedlung,* Ferdinand Enke Verlag, Stuttgart 1937, p. 46 and *6 Jahre,* p. 5 which gave 30,549 accepted from 61,197 applicants.

62. *Vierteljahrsheft,* op. cit., Note 28, Sec. III, p. 25 and p. 7.

63. Kummer to Darré 8 August 1934, BA NS26/946.

64. Borsig, 1934, op. cit., ch. 3, Note 15, p. 21 and letter 28 December 1934, HSAD 1073.

65. *Wirtschaftsblatt der Industrie und Handelskammer Berlin,* Heft 16-17, BA R22/2128.

66. Borsig, 1934, op. cit., ch. 3, Note 15, p. 21.

NOTES, *Chapter 11*

1. Domarus, 1965, op. cit., ch. 3, Note 38, p. 173 and 'Das Ziel', *Deutsche Agrarpolitik,* 1933, p. 5.

2. *Statistisches Jahrbuch,* 1941-42, p. 284 and ibid, 1936, p. 506, and correspondence 7 April 1933 and 6 June 1933, BA R43I/1461 and 1463.

3. Gesetz zum Schutz der deutschen Warenausfuhr, *RGB*(I), p. 667.

4. R. Erbe, *Die NS Wirtschaftspolitik 1933-39 in Lichte der modernen Theorie,* Polygraphischer Verlag, Zurich 1958, p. 100.

5. *Statistisches Jahrbuch,* 1938, p. 256 and FTEC, op. cit., p. 61.

6. Letter to Finance Minister 5 November 1934, BA R2/18018 and Dr Krohn of RNS Department II in an interview with the author.

7. Letter to Lammers (Chancellery) 13 January 1934, BA R43II/305c.

8. Correspondence and minutes, BA R43II/303b, 303c, etc.
9. Schacht/Darré correspondence April 1934, BA R43II/201.
10. Vide his letter to Lammers 24 February 1934, BA R43II/305c, and speech at Gauleiter Oberpräsident conference DNB 29 August 1935.
11. Letter to Ministry of Finance 30 November 1935, BA R2/18018 and *VB* 29 January 1935.
12. Borsig, 1934, op. cit., ch. 3, Note 15, p. 40.
13. *NS. Landpost,* 14 January 1935.
14. Letter to Goering 1 November 1936, ND 146.
15. OBF Langenberg 24 March 1935, HSAD 130.
16. Schacht, *Abrechnung mit Hitler,* Rohwohlt Verlag, Hamburg 1949, p. 49.
17. Schacht/Darré correspondence April-May 1934, BA R43II/201.
18. Regpräsident Hanover to Prussian Ministry of Agriculture 20 November 1934, NSA 122a XXXII 80, and Thyssen, 1958, op. cit., ch. 1, Note 23, p. 300.
19. FTEC p. 122 and *Statistisches Jahrbuch,, 1941-42,* p. 663.
20. REM to Lbschs. 1 November 1934, NSA 310 I B46, and NS Hago reports October-November 1935, ibid.
21. M. Broszat, 'The Concentration Camps 1933-1945', in Krausnick, 1968, op. cit., ch. 7, Note 19, p. 450.
22. Arbeitertum, 15 November and 1 December 1935.
23. W. Birkenfeld, 'Zur Entstehung und Gestalt des Vierjahrsplanes', in K. E. Born (Ed.), *Moderne deutsche Wirtschaftsgeschichte,* Kiepenheuer & Witsch, Cologne 1966, p. 411 and D. Petzina *Autarkiepolitik im Dritten Reich,* Stuttgart 1968, p. 33ff.
24. Memo ADC Schacht. It is not clear whether this was sent to anyone or merely represented his private views.
25. W. Treue, 'Hitlers Denkschrift über den 4 Jahresplan', *VJH* 1955, p. 194
26. Erbe, 1958, op. cit., Note 4, p. 100.
27. Petzina, 1967, op. cit., ch. 3, Note 27, pp. 31 and 46 and Birkenfeld, op. cit., Note 23, p. 411ff.
28. *Trial Brief,* p. 26.
29. *Mein Kampf,* p. 146.
30. J. von Stackelberg, 'Das Ergebnis der Landwirtschaft', *Wirtschafts-jahrbuch,* 1939, p. 27.
31. Cabinet minutes 1 December 1936, BA R43I/1465.
32. Petzina, 1967, op. cit., ch. 3, Note 27, p. 83.
33. Uebersicht über die zur Förderung der Landwirtschaft (einschliesslich Fischerei) in den Jahren 1933 bis 1938 verausgabten Beträge, BA R2/18019.
34. Letter to Ministry of Finance 21 April 1937, BA R2/18019.
35. Petzina, 1967, op. cit., ch. 3, Note 27, p. 86, *Frankfurter Zeitung* 22 July 1937, and *West Deutscher Beobachter* 8 December 1936.
36. *RMB* 3 March 1937.
37. Gesetz zur Sicherung der Landbewirtschaftung *RGB*(I) p. 442, and letter 16 July 1937, BA R2/18018.
38. Minute 10 October 1937, NSA 331 B35.
39. For example, correspondence 1938-39 HSAD Vohwinkel.

40. *Frankfurter Zeitung* 19 December 1937.
41. *50 Jahre*, op. cit., ch. 8, Note 20, p. 33.
42. Dr Krohn in an interview with the author.
43. Borsig, 1934, op. cit., ch. 3, Note 15, p. 33.
44. *Bauerntum,* published by the SS 1942, p. 49, and Bente, 1937, op. cit., ch. 3, Note 2, p. 105.
45. Arbeitertum, 1 October 1935, and NSA 122a XXXII 80.
46. K. Emig, 'Das Reichsministerium für Ernährung und Landwirtschaft', in *Das Dritte Reich im Aufbau,* Vol. IV, p. 292, *RMB* 13 and 26 March 1937, J. Müller *Deutsches Bauerntum zwischen gestern und morgen,* Würzburg 1939, p. 103, and *West Deutscher Beobachter* 24 March 1937.
47. von Hellerman, 1939, op. cit., ch. 10, Note 1, p. 63 and Schoenbaum, 1966, op. cit., ch. 5, Note 21, p. 13.
48. *Blut und Boden*, RNS Verlag, Berlin 1936, pp. 466 and 478.
49. *RMB* 28 May 1936, F. Sohn, 'Allgemeiner agrarpolitischer Bericht', *Berichte über Landwirtschaft,* 1939, p. 313, and H. von der Decken, 'Die Mechanisierung in der Landwirtschaft', *Vierteljahrsheft zur Konjunkturforschung,* 1938-39, pp. 355-56.
50. Gesetz über die Landbeschaffung für Zwecke der Wehrmacht and Gesetz über die Regelung des Landbedarfs der öffentlichen Hand, *RGB*(I) p. 467.
51. *Trial Brief,* p. 31, and correspondence, September 1934, BA R43I/1301 for peasant petitions against land-loss.
52. H. Reischle, 'Die deutsche Ernährungswirtschaft', in *Das Dritte Reich im Aufbau,* Vol. II, Quelle and Meyer, Leipzig, 1938, p. 282.
53. *Die landwirtschaftlichen Verhältnisse,* etc. p. 8 NSA 122a XXXII 88a.
54. 6 March 1939, HSAD 854.
55. Backe to Himmler 20 January 1943, ADC Willikens.
56. Petzina, 1967, op. cit., ch. 3, Note 27, p. 94ff, unless otherwise stated, for this section.
57. *Frankfurter Zeitung,* 21 February 1938.
58. FTEC, op. cit., ch. 6, Note 41, p. 122.
59. Ibid., pp. 124-25 and p. 178ff.
60. Compiled from *Statistisches Jahrbuch,* 1941-42, pp. 284, 436-7 and 663, ibid., 1936, pp. 350-51, Petzina, 1967, op. cit., ch. 3, Note 27, pp. 93-95, *Bauerntum,* 1942, op. cit., Note 43, p. 52, *Die Landesbauernschaften in Zahlen,* op. cit., ch. 10, Note 42, pp. 26-43, FTEC, op. cit., ch. 6, Note 41, p. 122ff., and M. Tracy, *Agriculture in Western Europe,* Frederick A. Praeger, New York, 1964, pp. 207-9.

NOTES, *Chapter 12*

1. *Die wirtschaftliche Lage der Landarbeiter und Landarbeiterinnen in Deutschland,* Berlin 1928 published by Agricultural Workers Union, p. 12.

2. von Hellermann, 1939, op. cit., ch. 10, Note 1, p. 18.
3. M. Brand, *Die Frau in der deutschen Landwirtschaft,* Franz Vahlen Verlag, Berlin 1939, p. 203.
4. *Statistisches Jahrbuch,* 1941-42, p. 18.
5. Seldte to Lammers 2 March 1933, R43II/213b.
6. Vide NSDAP Economic Policy Committee meeting 10 June 1934, BA NS 26/953.
7. *VB* 1 June 1934.
8. *WB* 5 July and 11 October 1933.
9. ADC Angelegenheit Ost Preussen 1933 September 1933.
10. *Statistisches Jahrbuch,* 1936, p. 325.
11. Ministry of Labour to Chancellery May 1934, BA R43I/1469.
12. *Frankfurter Zeitung* 5 October 1934 and Wunderlich, 1961, op. cit., ch. 7, Note 36, p. 295.
13. J. Scheuble, *Hundert Jahre staatliche Sozialpolitik 1839-1939,* W. Kohlhammer Verlag, Stuttgart 1957, p. 439.
14. 19 December 1935, NSA 122a 13 and Broszat, 'The Concentration Camps 1933-1945', in Krausink, 1968, op. cit., ch. 7, Note 19, p. 450.
15. Correspondence May-June 1936, HSAD 127.
16. Vide von Hellermann, 1939, op. cit., ch. 10, Note 1, p. 39.
17. See Westdeutsche Land Zeitung 28 April 1939.
18. Regpräsident report December 1937-January 1938 HSAD 1969.
19. Darré to all ministries 14 September 1940 BA R2/18197.
20. BA R/16 for this survey, based on eighteen districts.
21. Letters 26 January and 30 November 1935, BA R2/18018.
22. *Die Landesbauernschaften in Zahlen,* op. cit., ch. 10, Note 42, pp. 176-7, and von Hellermann, 1939, op. cit., ch. 10, Note 1, p. 38.
23. Müller, 1939, op. cit., ch. 11, Note 46, p. 92 and Brand, 1939, op. cit., Note 3, pp. 37, 44 and 123.
24. 'Stillstand in der Erzeugungsschlacht? ' *Odal* March 1939 p. 153.
25. *Auf dem Wege zur völkischen Schule* 1939 published by RNS and NS Lehrerbund p. 7.
26. von Hellermann, 1939, op. cit., ch. 10, Note 1, p. 31 and KBF to Lbsch. Bavaria 8 December 1937 BA R2/18291.
27. A. M. Koeppens *Das deutsche Landfrauenbuch,* RNS Verlag, Berlin 1937, p. 121.
28. Müller, 1939, op. cit., ch. 11, Note 46, p. 108 and *Die Betreuung des Dorfes* DAF 1937 p. 25.
29. Müller, ibid., p. 71, ADC Reichnährstand Wendt, 29 January 1938, Landrat Düren to Regpräsident Aachen 2 April 1937 and 7 June 1938 HSAD 1066 and 1070 and Trustee of Labour report January-February 1938 BA R43II/528.
30. Cited in Brand, 1939, op. cit., Note 3, p. 48.
31. Deputy Gauleiter E. Hanover to Hitler's Staff Office 25 January 1939 BA R43II/213b: Minute No. 6816 B 13 March 1939 makes the same point for Hessen-Nassau, ibid.
32. Dr H. Hajek (Ed.), *Landwirtschaftliche Berufe für Jungen,* Reichseltern-

warte Verlag, Berlin 1941, pp. 9-10 and E. Steiner *Agrarwirtschaft und Agrarpolitik,* 1939 University of Munich dissertation, p. 28.

33. Seldte to Lammers 16 March 1938 BA R43II/533.

34. Minute No. 4004 6 January 1939, HSAS 646, E/130 IV, Letter 16 January 1939 BA R43II/213b, and deputy Gauleiter in E. Hanover, see Note 31.

35. KBF Krumbach to Lbsch. 22 January 1938 BA R2/18291.

36. von Hellermann, 1939, op. cit., ch. 10, Note 1, p. 49.

37. Letter to Darré 29 January 1938 ADC Reichsnährstand Wendt.

38. Arbeitertum 15 December 1937, and K. Jacoby (Ed.), *Das Dritte Reich im Aufbau* Vol. I, Quelle and Meyer, Leipzig, 1938, p. 73.

39. Vide *Die Betreuung,* op. cit., Note 28, pp. 7-23, 'Arbeitertum' 1 April 1937, and von Hellermann, 1939, op. cit., ch. 10, Note 1, p. 31.

40. Bente, 1937, op. cit., ch. 3, Note 2, p. 172, *Bauerntum,* 1942, op. cit., ch. 11, Note 43, p. 45, *Statistisches Jahrbuch,* 1941-42, p. 359 and 613, and H. Reischle, 'Einkreisung der Unterwertung' *Odal* August 1939, p. 651.

41. According to Herr von Rohr in an interview with the author.

42. Ministry of Interior report 9 July 1935 BA R43II/317 and Chancellery minute 4 September 1935 BA R43II/318.

43. Correspondence February 1935 BA R43II/552, and Hess to Seldte 27 April 1935 BA R41/24.

44. Conference minutes 2 May 1935 and letter 12 May 1935 BA R2/18185.

45. Conference minutes 29 August 1935. Wholesale food prices went up 3.2 percent between June and 7 August 1935. Letter 29 August 1935, BA R43II/318.

46. Reischle to Bormann 14 March 1942 ADC Reischle.

47. See Chancellery minutes 19 and 25 May 1938 BA R43II/194.

48. BA R43II/213b.

49. Wochenblatt der Landesbauernschaft Schleswig-Holstein, 11 February 1939.

50. *Statistisches Jahrbuch*, 1941-42, p. 605. The figures given are for self-employed farmers, not for agriculture as a whole.

51. Reichsamt für Agrarpolitik report February 1938, HSAS 646 E 130/IV, and Seldte to Lammers 24 March 1938 BA R43II/533.

52. Müller, 1939, op. cit., ch. 11, Note 46, pp. 39 and 41.

53. 'Bemerkungen zur Erzeugungsschlacht' January 1938 BA R2/18291.

54. Ibid.

55. Reports and correspondence January-April 1938 and 3 March 1939 BA R43II/528 and 25 January 1939 BA R43II/213b.

56. WB 3 May 1933, and H. Müller 'Soziale Arbeit in der HJ' *Soziale Praxis* Heft 28 1934.

57. *Statistisches Jahrbuch,* 1936, p. 326.

58. von Hellermann, 1939, op. cit., ch. 10, Note 1, p. 42. Somewhat later Seldte gave a shortfall of 250,000 in a letter to Lammers 16 March 1938 BA R43II/533.

59. Brand, 1939, op. cit., Note 3, p. 121.

60. *Frankfurter Zeitung* 17 January 1938 and 26 February 1939, Labour Exchange Düsseldorf circular 24 February 1939, HSAD 26, and von Hellermann, 1939, op. cit., ch. 10, Note 1, p. 49.

61. Trustee of Labour report June 1937 BA R43II/528 and Ministry of Labour circular 10 May 1939 HSAD 26.

62. Deputy Gauleiter E. Hanover letter, see Note 31.

63. RMB 15 February 1938, and *Die Landesbauernschaften in Zahlen,* op. cit., ch. 10, Note 42, p. 174.

64. von Hellermann, 1939, op. cit., ch. 10, Note 1, p. 45, and KBF Niederberg to Lbsch. 12 March 1936 HSAD 127.

65. H. Kühne 'Der Arbeitseinsatz im Vierjahresplan', *Jahrbücher für National-ökonomie und Statistik,* 1937, p. 704, and von Hellermann, ibid., p. 44.

66. *Die Landesbauernschaften in Zahlen,* op. cit., ch. 10, Note 42, p. 52 and von Hellermann, ibid., p. 44.

67. Bente, 1937, op. cit., ch. 3, Note 2, p. 91 and Reichsamt für Agrarpolitik report January 1938.

68. Vide Landrat Geilenkirchen – Heimsberg to Regpräsident Aachen 25 June 1937 HSAD 1067, VB 2 February 1935, and KBF Niederberg to Lbsch. 28 March 1936 HSAD 127.

69. Landrat reports 5 August and 4 June 1937 HSAD 1067, Reichsamt für Agrarpolitik report October 1938 HSAS 646 E 130/IV, *Daily Telegraph* 23 July 1938, RMB 1 August 1939, and NSA 331 B35.

70. Correspondence January 1938 BA R2/18291.

71. Ministry of Labour letter 3 March 1939 BA R43II/528.

72. RGB(I), 1938, p. 652, *Frankfurter Zeitung* 21 January 1939 and Scheuble, 1957, op. cit., Note 13, pp. 434-5.

73. Deputy Gauleiter E. Hanover, see Note 31.

74. Ministry of Labour to all ministries/Chancellery BA R43II/528.

NOTES, *Chapter 13*

1. *WB* 2 and 9 August 1933.

2. Letter 16 September 1935 HSAD 1023.

3. KBF Euskirchen to Lbsch. Rhineland 6 November 1935 HSAD 1028.

4. Gesetz über die Feiertage *RGB*(I) p. 129 and *VB* 3 October 1933.

5. Ibid and Verhey, op. cit., ch. 1, Note 5, pp. 61-2.

6. *VB* 4 October 1937.

7. Richtlinien für die Beauftragten der Fachschaft für Brauchtum und Bauernkultur LBF Rhineland 7 March 1934 HSAD 1023.

8. Correspondence September-November 1937 BA R16/23, Lbsch. Kurhesse to Rhineland 10 November 1934 HSAD 1040 and KBF Cochem-Zell to Lbsch. 18 January 1937 HSAD 1051.

9. Kbsch. Euskirchen to Lbsch. 18 November 1936 Ibid.

10. Zeitungsdienst des RNS No. 154 13 July 1937, Lbsch. Rhineland 19 October 1935 HSAD 1051, and 9 November 1936 HSAD 1023.

11. Internal order 12 December 1934 HSAD 866.

12. Vide NS Landpost 20 May 1935.

13. 8 December 1936 HSAD 1023, and NS Landpost 16 October 1936.

14. Richtlinien für die Ehrung alteingesessener Bauerngeschlechter 13 December 1934 HSAD 866.

15. *RNS* Goslar to Lbsch. Rhineland, 16 July 1936, HSAD 1023.

16. Lbsch. Rhineland to Goslar, 5 July 1937, HSAD 866.

17. Correspondence June 1939, LUD 1625/K710.

18. 14 May 1935 LUD 1631/K710.

19. Correspondence March 1935 and February 1937 HSAD 1017.

20. Correspondence May 1936 HSAD 1058 and RNS Dienstnachrichten No. 15, 18 April 1936.

21. Wochenblatt der Lbsch. Württemberg, 18 May 1935.

22. Circular Kbsch. Rees and Dinslaken, January 1935, HSAD 1023.

23. Letter to Lammers, 7 March 1938, BA R4311/529.

24. Darré to LGFs, 22 August 1931, ND 142 and 30 January 1932, ND 145.

25. R. W. Darré 'Das Ziel' in H. Reischle (Ed.) *Deutsche Agrarpolitik,* 1934, op. cit., ch. 5, Note 3, pp. 14-15.

26. K. Verhey, 1965, op. cit., ch. 1, Note 5, p. 84.

27. *RNS* Goslar to REM, 12 August 1935, HSAS 653 E 130/IV.

28. Koeppens, 1937, op. cit., ch. 12, Note 27, p. 233.

29. Correspondence January-March 1939, NSA 331 B61.

30. Kbsch. Hameln-Bad Pyrmont letter, 26 October 1937. NSA 331 B50.

31. KBF Hameln-Bad Pyrmont 14 April 1934, NSA 331 B1, and correspondence 18 October 1934 and 21 August and 17 November 1939 ADC Reichsnährstand B.

32. Circular March 1937, BA R16/28 and letter 8 August 1935, HSAD 130.

33. Kbsch. Bussen to Lbsch. Württemberg, 19 May 1938, LUD 13/K710.

34. *WB* 5 July 1933.

35. Unterweisung in Volks- und Staatsbürgerkunde in den landwirtschaftlichen Lehranstalten *RMB*, 25 November 1933.

36. K. Verhey, 1965, op. cit., ch. 1, Note 5, pp. 85 and 88.

37. KBF Niederberg to Lbsch. Rhineland, 30 March 1936, describes the exam for three girl apprentices in these terms, HSAD 127.

38. Conference recommendations published as *Auf dem Wege zur völkischen Schule* RNS 1939.

39. W. Kinkelin, 'Bauerntum und SS' *Odal*, April 1936, p. 247.

40. Quoted in VB 11-12 February 1935.

41. 'Die Frau im Reichsnährstand' *Odal,* September 1934, p. 148.

42. 'Der Rheinische Bauernstand' 13 October 1934, HSAD 1140.

43. Details from *Die biologisch-erbbiologische Untersuchung der Erbhofbauern,* 1935 Bavarian Ministry of Interior Health Department pp. 5ff.

44. Correspondence December 1937, HSAD 1044 and August 1937, HSAD 854.

45. Lbsch. Saarland circular, October 1935, HSAD 1044.

46. Vide Kbsch. Bergheim letter, 1 March 1936, Ibid.

47. H. Gauch, *Die germanische Odal – oder Allodverfassung,* Blut und Boden Verlag, Berlin 1934, p. 6.

48. Minister of Justice Württemberg to regional Ministry of the Interior, 3 October 1934 and letter 20 June 1934, HSAS 321 E 130/IV.

49. Wochenblatt der Landesbauernschaft Württemberg, 25 November 1935, LUD 1625/K710, exhibition details LUD 1666/K710 and RNS Goslar to Lbsch. Rhineland, 16 July 1936, HSAD 866.

50. Cf. the sub-title of the *Origin of Species* itself, 'The preservation of favoured races in the struggle for life'.

51. *Statistisches Jahrbuch 1936* p. 11 and pp. 35-7, and C. W. Guillebaud, *The Social Policy of Nazi Germany,* CUP, London 1940, p. 101.

52. See Borsig, 1934, op. cit., ch. 3, Note 15, p. 82 for some relevant comparisons and *Statistisches Jahrbuch 1936* p. 38.

53. F. Bürgdorfer, *Der Geburtenrückgang und seine Bekämpfung,* R. Schoetz, Berlin, 1929.

54. H. Haushofer, *Ideengeschichte der Agrarwirtschaft und Agrarpolitik,* Vol. 2, Bayerischer Landwirtschaftsverlag. Munich 1958, p. 126.

55. Müller's views were summarised in *Die biologische Lage des Bauerntums* published by Archive for Population Science and Policy, 1938.

56. Correspondence December 1937 – January 1938, ADC Reichsnährstand Martin Wendt, Wagner and Willikens.

57. Report 2 February 1939, HSAD 854.

58. Regional government conference 20 September 1933, HSAS 343 E130/IV.

59. This investigation, carried out in Württemberg by a sociologist named Münzinger is cited in the RNS report of 2 February 1939.

60. Kbsch. Hof-Saale to Lbsch. January 1938 BA R2/18291.

61. Letter to Darré, 29 January 1938, ADC Reichsnährstand Wagner.

62. 'Race' in this context meant Nordic, Alpine, Slav, Mediterranean or Falian, the five races allegedly comprising the German people: K. Kummer 'Der Bauer im Staat' in W. Clauss (Ed.) *Der Bauer in Umbruch der Zeit,* RNS Verlag, Berlin 1935, p. 54.

63. Vide Darré's article 'Geburtminderung durch das Erbhofgesetz? ', 1934, op. cit., ch. 8, Note 12.

64. In an article 'Das Deutsche Siedlungswerk' *Zentralblatt der Bauverwaltung* Heft 20, 16 May 1934, BA R43II/211.

NOTES, *Chapter 14*

1. Vide W. Meinhold, *Die landwirtschaftliche Erzeugungsbedingungen im Kriege,* Jena Gustav Fischer Verlag 1941, p. 18f.

2. Dr. H. Dommasck, 'Die Verbrauchsregelung in der deutschen Kriegsernährungswirtschaft', *Berichte über Landwirtschaft,* Vol. 28, 1943, p. 518f, H. Reischle, *Kann man Deutschland aushungern?* Berlin Zentralverlag der NSDAP 1940, p. 17 and R. W. Darré, *Der Schweinemord,* Munich 1937.

3. H. Backe, *Um die Nahrungsfreiheit Europas,* Leipzig, 1942, p. 151. A. Dallin, *German Rule in Russia 1941-1945,* Macmillan 1957, p. 308 gives a figure of seven million tons.

4. Verordnung über die öffentliche Bewirtschaftung landwirtschaftlicher Erzeugnisse, *RMB* 1939, p. 907, and Verordnung zur vorläufigen Sicherstellung des lebenswichtigen Bedarfs etc. *RGB*(I) p. 1498.

5. *50 Jahre,* op. cit., ch. 8, Note 20, p. 34 and *RMB* 1939, pp. 1181-2.

6. Reischle, 1940, op. cit., Note 2, p. 55 and K. Backhaus, 'Warum Malthus nicht recht behielt', *Der Vierjahresplan,* February 1944, p. 36f.

7. *Statistisches Handbuch von Deutschland,* Munich 1949, p. 402.

8. 'Die Ernährungslage hüben und drüben', *Der Vierjahresplan,* 15th December 1944, p. 414f and Backe, op. cit., Note 3, for comparative yields.

9. Meinhold 1941, op. cit., Note 1, p. 89, T. Thyssen, 1958, op. cit., ch. 1, Note 23, p. 289 and Anordnung etc., *RMB,* 16 February 1942, p. 138.

10. 'Erntezeit', *Der Vierjahresplan,* 15 August 1943, p. 295 and *Der Aktivist* published by Gau S. Hanover-Brunswick, No. 4, 1944, p. 41.

11. E. Woermann, 'Europäische Nahrungswirtschaft'. *Nova Acta Leopoldina,* Vol. 14, No. 19. Halle 1944, p. 13.

12. *Statistisches Handbuch* p. 190, and Reischle, 1940, op. cit., Note 2, p. 47.

13. C. A. Wegener, 'Die europäische Milch- und Fettwirtschaft', *Berichte über Landwirtschaft,* Vol. 28, 1943, p. 469.

14. Reischle, 1940, op. cit., Note 2, p. 48 and Thyssen, 1958, op. cit., ch. 1, Note 23, p. 290.

15. 'Die Bäuerin im totalen Kriegseinsatz'. *Der Vierjahresplan,* 15 August, 1943, p. 280f. See also *Bauerntum,* 1942, op. cit., ch. 11, Note 43, p. 24.

16. H. L. Fensch, 'Die Unterbewegung der Landarbeit in den verschiedenen Betriebsgruppen der Landwirtschaft', *Berichte über Landwirtschaft,* Vol. 28, 1943.

17. Backe, op. cit., Note 3, p. 149.

18. Anordnung zur Durchführung der Milcherzeugungsschlacht vom 6 März 1940, RMB 1940, pp. 217f and 1941, p. 743.

19. See Haushofer, 1963, op. cit., ch. 4, Note 33, p. 268 for the struggle needed to get the increase.

20. The actual premium varied during the war, the basis here is that of 1943. RMB 1943, p. 83: See F. Wunderlich, 1961, op. cit., ch. 7, Note 36, p. 196.

21. *Statistisches Handbuch,* p. 461. Vide also T. Thyssen, 1958, op. cit., ch. 1, Note 23, p. 289.

22. 'Die Ernährungslage hüben und drüben' 1944, op. cit., Note 8, p. 414, Backe, op. cit., Note 3, p. 160, and *Der Aktivist,* 1944, op. cit., Note 10, p. 40.

23. *Statistisches Handbuch,* p. 190.

24. Anordnung . . . zur Durchführung der Schlachtvieherzeugung, RMB 1942, p. 138, and *Statistisches Handbuch,* p. 461, 124-6 and 190.

25. Meinhold, 1941, op. cit., Note 1, pp. 89-90, Massnahmen gegen die Verfutterung von Kopfkohl etc. RMB 1943, p. 935. Hausschlachtungen etc., ibid., pp. 278 and 586.

26. Vide *50 Jahre,* op. cit., ch. 8, Note 20, p. 37.

27. RGB(I) 1940, p. 610, ibid., 1942, p. 443 and ibid., 1943, p. 576 for these measures.

28. *Der Aktivist,* 1944, op. cit., Note 10, p. 36.
29. H. Boberach Ed. *Meldungen aus dem Reich,* Luchterhand Neuwied, 1965, pp. 285 and 476-7.
30. Stillegung von Handelsbetrieben in der Ernährungswirtschaft RMB 1943, p. 280f and ibid., p. 690.
31. Meinhold, 1941, op. cit., Note 1, p. 39 and 102, and *Statistisches Handbuch*, p. 187.
32. 'Die Erhaltung der Bodenfruchtbarkeit im Kriege', *Der Vierjahresplan,* March 1944, p. 62f.
33. Wunderlich, 1961, op. cit., ch. 7, Note 36, p. 188, and RMB 1942, p. 184.
34. In 1942 agricultural machinery output was 293,120 tons, well below the 1938 figure. Two years later it fell to 228,000, about half the 1938 production. *Statistisches Handbuch,* p. 301.
35. Haushofer, 1963, op. cit., ch. 4, Note 33, p. 270, and Dallin, 1957, op. cit., Note 3, p. 365.
36. Meinhold, 1941, op. cit., Note 1, p. 54, Haushofer, ibid., p. 268, and Boberach, 1965, op. cit., Note 29, p. 466.
37. Verordnung über die Beschränkung des Arbeitsplatzwechsels RGB(I) p. 1685.
38. Meinhold, 1941, op. cit., Note 1, p. 55, RGB(I) pp. 179-180 and Verordnung über den Einsatz zusätzlicher Arbeitskräfte etc. Ibid., p. 105.
39. RMB 1943, p. 175.
40. 'Die Bäuerin im totalen Kriegseinsatz' *Der Vierjahresplan,* 15 August 1943, p. 280.
41. Meinhold, 1941, op. cit., Note 1, pp. 66-7, and A. König, 'Die landw. (sic) Berufs- und Fachschulen im Dienste des Ernährungswerk', *Die deutsche Berufserziehung,* Ausgabe C. 15 June, 1942.
42. Verordnung über die Durchführung der Reichsarbeitsdienstpflicht für die weibliche Jugend RGB(I) p. 1693, ibid., 1941, p. 463 and Verordnung über die Dauer der Dienstzeit etc., Ibid., 1942, p. 336.
43. Vide *Arbeitsmaiden in der Nordmark* pub. by RAD area office, Kiel, 1943, pp. 7 and 70.
44. Haushofer, 1963, op. cit., ch. 4, Note 33, p. 269 and *Statistisches Handbuch,* p. 481.
45. *Der Aktivist,* 1944, op. cit., Note 10, p. 12 and Haushofer, ibid., p. 269.
46. Meinhold, 1941, op. cit., Note 1, p. 61.
47. Einsatz von Kriegsgefangenen bei Meliorationen etc. RMB 1939, pp. 1029-1030 and Ibid., 1940, p. 851.
48. Vide Boberach, 1965, op. cit., Note 29, pp. 309ff.
49. Ibid., p. 504 and *Rassenpolitik im Kriege*, published by Rassenpolitisches Amt. Gau South Hanover-Brunswick 1941. pp. 115-6 and 26.
50. J. Jeurink, 'Berufswahl und Abwanderung der Landjugend im Kreise Göttingen', *Die deutsche Berufserziehung*, 15 November 1942, and Wunderlich, 1961, op. cit., ch. 7, Note 36, p. 309.
51. For example, see *Die deutsche Berufserziehung*, ibid., Aug-Sept. 1942, p. 141 for a reference to such duties in the Warthegau.
52. Meinhold, 1941, op. cit., Note 1, p. 35.

53. *Statistisches Handbuch,* pp. 124-6 and p. 490.
54. See Boberach, 1965, op. cit., Note 29, p. 505 for the latter point.
55. *Statistisches Handbuch,* pp. 124-6.
56. Wegener, op. cit., Note 13, p. 462f for the figures.
57. Reischle, 1940, op. cit., Note 2, p. 80.
58. K. Brandt and others, *The management of Germany's agricultural food policies in World War II,* Vol. II. Stanford University Press, 1953, pp. 57 and 621f. Economic Staff East was the *ad hoc* body installed to plan the material exploitation of the Soviet Union.
59. For a detailed description of the incompetence and political infighting, Dallin, 1957, op. cit., Note 3, especially pp. 310f and p. 368.
60. Adapted from Brandt, 1953, op. cit., Note 58.
61. Ibid., pp. 610-13 and *Statistisches Handbuch,* p. 402ff.
62. Ibid., pp. 392f and Brandt, 1953, op. cit., Note 58, p. 273. Norway was also a net debtor for fats and had to be helped out by the Reich after 1941. Ibid., p. 611.
63. Vide E. Woermann, 1944, op. cit., Note 11.
64. *50 Jahre,* op. cit., ch. 8, Note 20, p. 38.
65. H. Schmitz, *Die Bewirtschaftung der Nahrungsmittel und Verbrauchsgüter 1939-1950,* Essen 1956, p. 357ff.
66. For the new organisation Koehl *RKFDV,* pp. 50f and Verordnung über die öffentliche Bewirtschaftung etc. RGB(I) 12 February 1940, p. 355.
67. Grundsätze und Richtlinien für den ländlichen Aufbau in den neuen Ostgebieten, *Landvolk im Werden,* K. Meyer (Ed.), p. 361.
68. J. Umlauf, 'Der ländliche Siedlungsaufbau in den neuen Ostgebieten', Ibid., p. 273f.
69. W. Zoch, 'Blickwendung zum Osten', ibid., p. 56f., *Bauerntum,* 1942, op. cit., ch. 11, Note 43, introduction and especially Chapter 8, where again it is maintained that only German peasants can make a land *germanisiert,* and *Rassenpolitik* published by SS, (probably in 1942) p. 53.
70. *Bauerntum,* ibid., p. 90-92. Himmler was honorary leader of the Land Service.
71. *Unser Dienst im September 1940,* issue for the new units in the east and south-east, pp. 14-27.
72. May 1941, 'Die Ostpolitik der staufischen Könige'.
73. Wunderlich, 1961, op. cit., ch. 7, Note 36, p. 314.
74. See *Die deutsche Volkschule,* published by the National Socialist Teachers League, January 1942, Heft 1, and Heft 4, April 1942, p. 111.
75. Quoted in *Die deutsche Berufserziehung,* 15 December 1942.
76. H. Lange, 'Zur Frage der Berufslenkung unserer Landjugend', Ibid 15 April 1942 and G. Stenkhoff, 'Landwirtschaftliche Berufs- und Fachschulen im Dienste der Nachwuchssicherung', Ibid 15 November 1942.
77. *Bauerntum,* 1942, op. cit., ch. 11, Note 43, p. 188 and H. Dittmar, *Der Bauer im Grossdeutschen Reich.* Deutscher Verlag Berlin, 1940, p. 21.
78. Zusammenarbeit zwischen SA und Reichsnährstand *RMB* 1941, p. 1043.
79. Dallin, 1957, op. cit., Note 3, p. 285.
80. Richtlinien für die Auswahl und Vermittlung neuer Bauern RMB 1941, p.

93.

81. F. Schlegelberger, quoted in *Der Prozess gegen die Hauptkriegsverbrecher*, Vol. XX, Nuremberg 1947, pp. 289f.

82. Quoted in Gies, 1967, op. cit., ch. 2, Note 10, p. 349.

83. Trial Brief, p. 223.

84. F. Harmening, in *Der Prozess,* op. cit., Note 81, Vol. XXI, p. 381.

85. L. Lochner, Ed., *The Goebbels Diaries,* p. 165ff.

86. IfZ 1622, Dr. Reischle in a personal interview and *50 Jahre*, op. cit., ch. 8, Note 20, p. 37.

NOTES, *Conclusion*

1. Reichsamt für Agrarpolitik January 1938, HSAS 646 R130/IV.

2. N. Rich, *Hitler's War Aims,* London 1974, p. 39.

3. Vide Guillebaud, 1939, op. cit., ch. 6, Note 23.

4. Thyssen, 1958, op. cit., ch. 1, Note 23, p. 299.

5. B. Vollmer, 1957, op. cit., ch. 6, Note 30, p. 115, and Ministry of Labour to ministries/Chancellery 3 March 1939, BA R43II/528.

6. Ibid.

7. Vollmer, ibid, p. 177f.

8. Bürgermeister Büderich to Lbsch. 25 April 1935, HSAD 1023.

9. Private letter 1 March 1937, ADC Reichsnährstand Darré.

10. Letter to Lammers, 7 March 1938, BA R43II/529.

11. Correspondence November 1934, BA NS26/955.

12. E. N. Peterson *The limits of Hitler's Power,* Princeton 1969, pp. 109-110.

13. Quoted in *Landvolk im Werden,* p. 271.

14. Backe, 1942, op. cit., ch. 14, Note 3, p. 197.

15. For figures of German rearmament, T. W. Mason in E. M. Robertson (Ed.) *The Origins of the Second World War,* Macmillan 1971, p. 116.

16. Brand, 1939, op. cit., ch. 12, Note 3, p. 117.

17. Zuschüsse zur Ansetzung für mittellose Neubauern RMB 1939, p. 1058.

18. In Reischle (Ed.), 1934, op. cit., ch. 5, Note 3, p. 72.

19. R. Cecil, *The Myth of the Master Race: Alfred Rosenberg and Nazi Ideology,* B. T. Batsford Ltd. 1972, p. 191.

20. For a fuller account, K. D. Barkin, *The Controversy over German Industrialisation 1890-1902,* Chicago U.P. 1970.

21. According to Dr. Reischle, the then head of the RNS staff office, in an interview.

22. For example, H. Reischle 'Von Ruhland zu Darré', *Die Landware* 14 September 1941.

23. His chief work was *Agrar- und Industriestaat* 1902.

24. K. Lange, 'Der Terminus "Lebensraum" in Hitler's Mein Kampf', *VJH* 1965.

25. Ibid., p. 428.
26. Ibid., p. 433.
27. See F. Fischer *Griff nach der Weltmacht,* Droste Verlag-Düsseldorf, 1967, pp. 104-5, 138-9, 142-3 and 234-5.
28. 'Grundsätze einer lebensgesetzlichen Agrarpolitik' in H. Reischle (Ed.) *Deutsche Agrarpolitik* 1934, p. 65.
29. Foreword by Darré to H. Backe *Das Ende des Liberalismus in der Wirtschaft.* RNS Verlag, Berlin 1942, 1938.
30. W. Abel *Agrarpolitik* Göttingen 1967, p. 22ff.
31. F. Schmidt, foreword to R. W. Darré *Neuordnung unseres Denkens,* Goslar 1941.
32. By 1936 the value of public contracts to the construction industry was 68 percent of total orders. Seldte to Lammers 22 October 1937, BA R43II/533.
33. *Trial Brief,* pp.14-16.
34. Fritzsche, quoted in Schoenbaum, 1966, op. cit., ch. 5, Note 21, p. 43.
35. Cecil, 1972, op. cit., Note 19, p. 116.
36. Backe, 1942, op. cit., ch. 14, Note 3, p. 220ff.
37. Quoted in R. Koehl, 1957, op. cit., ch. 10, Note 48, pp. 46-7.
38. Reprinted as 'Die Marktordnung der NS Agrarpolitik als Schrittmacher einer neuen europäischen Aussenhandelsordnung'. *Um Blut und Boden,* p. 511ff.
39. Backe, 1942, op. cit., ch. 14, Note 3.
40. Cited in K. Brandt, 1953, op. cit., ch. 14, Note 58, pp. 57-8.
41. Dr. A. Walter 'Landwirtschaftliche Zusammenarbeit in Europa'. *Berichte über Landwirtschaft,* Vol. 28, 1943.
42. Dallin, 1957, op. cit., ch. 14, Note 3, p. 305.
43. Brandt, 1953, op. cit., ch. 14, Note 58, p. 629.
44. Dallin, 1957, op. cit., ch. 14, Note 3, p. 287.
45. BA R43II/213 b.
46. Backe, 1942, op. cit., ch. 14, Note 3, p. 20.
47. Brandt, 1953, op. cit., ch. 14, Note 58, pp. 33 and 50-1.
48. E. M. Robertson, *Hitler's Pre-War Policy and Military Plans 1933-1939,* London 1963, Longmans, p. 110.
49. *Anweisungen für Propagandisten und Landwirtschaftsführer zum Agrarerlass,* Ostministerium 1942 (copy in BM).
50. H. C. Meyer, *Mitteleuropa in German Thought and Action.* Mouton. The Hague 1955, p. 86.

INDEX